This series aims to report new developments in physical research and teaching — quickly, informally, and at a high level. The type of material considered for publication includes:

1. Preliminary drafts of original papers and monographs

2. Lectures on a new field, or presenting a new angle on a classical field

3. collections of seminar papers

4. Reports of meetings

Texts which are out of print but still in demand may also be considered if they fall within these categories.

The timeliness of a manuscript is more important than its form, which may be unfinished or tentative. Thus, in some instances, proofs may be merely outlined and results presented which have been or will later be published elsewhere.

Publication of *Lecture Notes* is intended as a service to the international physical community, in that a commercial publisher, Springer-Verlag, can offer a wider distribution to documents which would otherwise have a restricted readership. Once published and copyrighted, they can be documented in the scientific libraries.

Manuscripts
Manuscripts are reproduced by a photographic process; they must therefore be typed with extreme care. Symbols not on the typewriter should be inserted by hand in indelible black ink. Corrections to the typescript should be made by sticking the amended text over the old one, or by obliterating errors with white correcting fluid. The figures (in the original size) ready for reproduction should be inserted into the text. Should the text, or any part of it, have to be retyped, the author will be reimbursed upon publication of the volume. Authors receive 50 free copies.

The typescript is reduced slightly in size during reproduction, therefore a large size of type should be used; best results will not be obtained unless the text on any one page is kept within the overall limit of 18 x 26.5 cm (7 x 10½ inches). The publishers will be pleased to supply on request special stationery with the typing area outlined.

Manuscripts in English, German or French should be sent to Springer-Verlag, 6900 Heidelberg, Postfach 1780.

Die *„Lecture Notes"* sollen rasch und informell, aber auf hohem Niveau, über neue Entwicklungen in der Physik berichten. Zur Veröffentlichung kommen:

1. Vorläufige Fassungen von Originalarbeiten und Monographien.

2. Spezielle Vorlesungen über ein neues Gebiet oder ein klassisches Gebiet in neuer Betrachtungsweise.

3. Seminarausarbeitungen.

4. Vorträge von Tagungen.

Ferner kommen auch ältere vergriffene spezielle Vorlesungen, Seminare und Berichte in Frage, wenn nach ihnen eine anhaltende Nachfrage besteht.

Die Beiträge dürfen im Interesse einer größeren Aktualität durchaus den Charakter des Unfertigen und Vorläufigen haben. Sie brauchen Beweise unter Umständen nur zu skizzieren und dürfen auch Ergebnisse enthalten, die in ähnlicher Form schon erschienen sind oder später erscheinen sollen.

Die Herausgabe der *„Lecture Notes"* Serie durch den Springer-Verlag stellt eine Dienstleistung an die physikalischen Institute dar, indem der Springer-Verlag für ausreichende Lagerhaltung sorgt und einen großen internationalen Kreis von Interessenten erfassen kann. Durch Anzeigen in Fachzeitschriften, Aufnahme in Kataloge und durch Anmeldung zum Copyright sowie durch die Versendung von Besprechungsexemplaren wird eine lückenlose Dokumentation in den wissenschaftlichen Bibliotheken ermöglicht.

Lecture Notes in Physics

Edited by J. Ehlers, München, K. Hepp, Zürich and
H. A. Weidenmüller, Heidelberg
Managing Editor: W. Beiglböck, Heidelberg

18

Proceedings of the Third International Conference on Numerical Methods in Fluid Mechanics

Vol. I
General Lectures. Fundamental Numerical Techniques

July 3-7, 1972
Universities of Paris VI and XI
Edited by Henri Cabannes and Roger Temam

Springer-Verlag Berlin Heidelberg GmbH 1973

ISBN 978-3-540-06170-0   ISBN 978-3-540-38377-2 (eBook)
DOI 10.1007/978-3-540-38377-2

Editors' Preface

This issue of Lecture Notes in Physics is the first of two volumes
constituting the Proceedings of the Third International Conference on
Numerical Methods in Fluid Mechanics, which was held at the University
of Paris VI, from July 3 to 7, 1972. Three general lectures and forty
eight short individual communications were presented at this conference;
the complete proceedings are published here. The general lectures
were given by Professor A. DORODNICYN, Director of the Computing Center
of the Academy of Sciences of the Soviet Union, who presented the
Soviet works dealing with the solution of Navier-Stokes equations; by
P. MOREL, professor at the University of Paris VI and Director at the
Laboratory of Dynamical Meteorology of the National Center of scientific research (C.N.R.S.), who presented the Problems of numerical simulation of geophysical flows; by Professor R.D. RICHTMYER of the University of Colorado, U.S.A., who spoke on Methods for (generally
unsteady) Flows with Shocks.

The individual communications have been separated into two groups:
Fundamental Numerical Techniques and Problems of Fluid Mechanics; in
each group they are published in the alphabetic order of the author,
or of the first of the authors.

Volume I contains the three general lectures and the thirteen communications on Fundamental Numerical Techniques. Volume II contains the
thirty five communications on Problems of Fluid Mechanics.

This Conference follows the conferences with the same topic hold at
Novossibirsk, U.S.S.R. in 1969, and at Berkeley, U.S.A. in 1970 (the
proceedings of which appeared in Lecture Notes in Physics, Vol. 8).
The French Organizing Committee was sponsored by Commissariat à
l'Energie Atomique, Electricité de France, Union des Chambres Syndi-
cales des Industries de Pétrole, in France, and also by the Office of
Naval Research and Air Force Office of Scientific Research, in the
U.S.A. The Universities of Paris VI and Paris XI, and the Centre
National de la Recherche Scientifique also helped the Committee in a
much appreciated manner.

We wish to thank all the persons who contributed to the success of the Conference, the participants for their scientific contributions, our colleagues and younger researchers for their help in the organization and Mrs. M.T. CARTIER and Miss S. DELABEYE for their excellent secretarial work.

Finally we wish to express our appreciation to Dr. W. BEIGLBÖCK and the Springer-Verlag Company for the rapid publication of these proceedings in the series of Lecture Notes in Physics.

January 25, 1973   HENRI CABANNES
                                                        ROGER TEMAM

Contents

Volume II

REVIEW OF METHODS FOR SOLVING THE NAVIER-STOKES EQUATIONS

A. A. Dorodnicyn

During the last 10 - 12 years the numerical solution of the complete equations
of motion of a viscous fluid - the Navier-Stokes equations - has been investigated
intensively.

Of course, interest in the Navier-Stokes equations arose much earlier. Although the theory of the boundary layer permits us to estimate the influence of
viscosity in many practically important cases (perhaps in most of them), nevertheless some phenomena, important in practice, are not described by boundary layer
theory. Among them must firstly be mentioned flow with separation. This phenomenon
occurs sometimes in flows around ships, planes, or rockets in spite of all possible
means being used to avoid it.

It is to be noted that in many complicated cases we cannot imagine the picture
of viscous flow even qualitatively and numerical calculations sometimes reveal quite
unexpected features of motion.

Finally there is the problem of turbulence. I don't think that complete information on the structure of the turbulence - one which will permit us to construct
an adequate theory of turbulance - can ever be derived from experimental observations.

The main difficulty here consists in the non-local character of the relations
between the stress-tensor and the deformation tensor. The non-linearity of this
relation seems to introduce less complication.

I think the numerical solution of the Navier-Stokes equations is the only way
to obtain sufficient information and the application of the methods of mathematical
simulation to this information will possibly bring us to some mathematical models
of turbulence with good precision.

In fact, modern computers are not yet sufficiently effective for solving the
Navier-Stokes equations with values of Reynolds numbers corresponding to fully
developed turbulence.

But even "pessimistic" forecasts of progress in computers gives us the hope
that in some 10 - 15 years such a possibility will become quite real.

The beginning of the work on the numerical calculations of the viscous fluid
motion was caused by the state of computers which just 10 or 12 years ago achieved
the level permitting a start to the solution of the problem.

Nevertheless the problem of solution of the Navier-Stokes equations continues
to be one of the most difficult problems in the mechanics of a continuous medium
and it always requires the fullest possible use of computers.

I. The Methods of Solution

Two properties of the Navier-Stokes equations are the main source of difficulty in their numerical solution: 1. high order of the system (order 4 in the
simplest case of an incompressible fluid), and 2. the unbounded domain of the
solution (for the most practically interesting cases), together with the elliptical
character of the equations.

When the Reynolds number is large a third difficulty appears - strong irregularity in the distribution of physical values (velocity, temperature, etc.).

As the system of Navier-Stokes equations is nonlinear its solution can be found only by means of some iterational procedure. It is well known in numerical analysis that it is advantageous to work with differential equations of the second order - since to solve them the numerical procedure can be reduced to the solution of a three-diagonal algebraic system of equations, and this means the number of necessary arithmetical operations falls from an order $O(n^3)$ to an order $O(n)$, where n is the number of unknown values.

It is quite natural, therefore, to try to construct the iterational procedure in such a way that a sequence of separated second-ordered differential equations has to be solved at each step of the iteration. For this purpose we always used the modification of boundary conditions on solid surfaces which gives the values of the stream-function and vorticity on boundaries at each step of the iteration.

To make the idea clear we consider the plane motion of an incompressible fluid.

For the solution of steady-flow problems three methods were used; to each of them a different computational procedure can be applied.

1. Method of successive approximations

In this method the steady system of Navier-Stokes equations is solved directly.

The system can be written in the form

$$\Delta\omega - 2\frac{\partial\omega}{\partial\xi} = 2\left[\frac{\partial\phi}{\partial\eta}\cdot\frac{\partial\omega}{\partial\xi} - \frac{\partial\phi}{\partial\xi}\cdot\frac{\partial\omega}{\partial\eta}\right] \tag{I}$$

$$\Delta\phi = F(\xi,\eta)\omega$$

where ξ,η - are some "canonical" coordinates (usually, the potential and stream-function in ideal flow), ω - dimensionless vorticity, ϕ - dimensionless additional stream-function (the full stream-function $\psi = \phi+\eta$).

On solid surfaces, the equations of which in canonical coordinates have the form: $\eta = $ const, the real boundary conditions are

$$\phi = \text{const}, \quad \frac{\partial\phi}{\partial\eta} = -1 \tag{II}$$

For reducing the system (I) to a sequence of second-order equations we modify the second boundary condition from (II)

$$\omega = \alpha(\xi)\left[1 + \frac{\partial\phi}{\partial\eta}\right] + \omega$$

and construct successive approximations in the following way

$$\Delta\omega_{n+1} - 2\frac{\partial\omega_{n+1}}{\partial\xi} = 2\left[\frac{\partial\phi_n}{\partial\eta}\cdot\frac{\partial\omega_n}{\partial\xi} - \frac{\partial\phi_n}{\partial\xi}\cdot\frac{\partial\omega_n}{\partial\eta}\right]$$

$$\Delta\phi_{n+1} = F(\xi,\eta)\omega_{n+1} \tag{III}$$

$$\phi_{n+1} = \text{const}, \quad \omega_{n+1} = \alpha(\xi)\left(1 + \frac{\partial\phi_n}{\partial\eta}\right) + \omega_n$$

So we see that at each step of the iteration two separated equations of the second order are to be solved.

The coefficient $\alpha(\xi)$ in the boundary conditions (we call it the "relaxation parameter") is introduced to secure the convergence of successive approximations.

If these approximations are convergent the real boundary condition (second in (II)) will automatically be satisfied.

2. "Real" stabilization method

The solution of the steady motion problem is obtained in this method as the limit of the unsteady Navier-Stokes equations when time $t \to \infty$.

The system of equations in canonical coordinates can be written

$$F(\xi,\eta) \frac{\partial \omega}{\partial t} = \Delta\omega - 2 \left(\frac{\partial \psi}{\partial \eta} \cdot \frac{\partial \omega}{\partial \xi} - \frac{\partial \psi}{\partial \xi} \cdot \frac{\partial \omega}{\partial \eta} \right)$$

$$\Delta\psi = F(\xi,\eta)\omega \tag{IV}$$

(with corresponding choice of time unit).

This system is used naturally for the solution of real unsteady problems, but when we are interested only in finding the final steady solution, the boundary conditions on the solid surface can be modified in such a way that for each time-step the values of vorticity (ω) and stream function (ψ) will be known and equations for ω and ψ become separated.

This modified boundary condition is

$$\frac{\partial \omega}{\partial t} = \alpha(\xi,t) \frac{\partial \psi}{\partial \eta} \qquad (\eta = \text{const}) \tag{V}$$

Obviously, the real condition $\partial\psi/\partial\eta = 0$ will be satisfied when the process tends to a steady state.

The real nonstationary system of Navier-Stokes equations is not convenient for numerical calculation, as it is a combination of parabolical and elliptical equations. When the steady state only is of interest we are not obliged to use this real system. Any nonstationary system can be used with the same steady part although perhaps it has no physical sense. So to avoid the bad properties of the real system we use the third method called:

3. "Artificial" stabilization method

In this method the solution of the steady problem is obtained as the limit $(t \to \infty)$ of the solution of the system of parabolic equations:

$$\frac{\partial \omega}{\partial t} = \Delta\omega - 2 \left(\frac{\partial \psi}{\partial \eta} \cdot \frac{\partial \omega}{\partial \xi} - \frac{\partial \psi}{\partial \xi} \cdot \frac{\partial \omega}{\partial \eta} \right)$$

$$\frac{\partial \psi}{\partial t} = \Delta\psi - F(\xi,\eta)\omega \tag{VI}$$

The boundary condition on a solid wall has the same form (V).

II. The Computational Procedure

In the methods of successive approximation, as we have seen, the problem is reduced to successive solutions of the Helmholz and Poisson equations. There are many effective methods for the numerical solution of these classical equations, especially taking into account that the forms of the domains in canonical coordinates are very simple (strips or planes with sections).

In our calculation we usually used the matrix factorization method. This is well known too, so it is not necessary to describe it here.

As the most interesting problems connected with the Navier-Stokes equations deal with flow in an infinite domain, the question arises how to transmit the condition from infinity to some finite distance.

In the case of the flow in a channel with parallel walls at infinity the boundary conditions for sufficiently large ξ can be written in the form:

$$\frac{\partial \omega}{\partial \xi} = 0 \quad , \quad \frac{\partial \psi}{\partial \xi} = 0 \qquad \text{when} \quad |\xi| = X_o \gg 1$$

As the flow in this case tends to Poiseuille flow the comparison with it can serve as some kind of control of accuracy.

When the asymptotic expression for the solution can be established (as, for instance, in the case of a semi-infinite flat plate), the boundary conditions at a large but finite distance can be deduced from these expressions. For example, in the flow around the flat plate the asymptotic expressions are:
for $\xi \gg 1$

$$\omega \sim \frac{1}{\sqrt{x}} \quad , \quad \phi \sim \sqrt{x}$$

Hence

$$\frac{\partial \omega}{\partial x} \cong - \frac{\omega}{2x} \quad , \quad \frac{\partial \phi}{\partial x} \cong \frac{\phi}{2x}$$

and the boundary condition taken for calculation was:
when $\xi = X_o \gg 1$

$$\frac{\partial \omega}{\partial x} = - \frac{\omega}{2 X_o} \quad , \quad \frac{\partial \phi}{\partial x} = \frac{\phi}{2 X_o}$$

Again the well known Blasius solution for the boundary layer flow past a flat plate could be used as a control of the accuracy of the calculations.

In the problems of external flow around some finite body, far from this body the solution of the Navier-Stokes equations asymptotically tends to the solution of the Oseen equations. This asymptotic equation gives the relation between the values of normal derivatives ω or ϕ and these functions themselves. In finite-difference approximation this relation will have the form:

$$\frac{\partial \vec{\omega}}{\partial n} = A\vec{\omega} \quad , \quad \frac{\partial \vec{\phi}}{\partial n} = B\vec{\phi} + C\vec{\omega}$$

Here by the "vector" $\partial\vec{\omega}/\partial n$ we understand the set of values of $\partial\omega/\partial n$, $\vec{\omega}$ - the set of values of ω on the boundary, chosen far enough from the body. The matrices A, B, and C are calculated from the general solution of the Oseen equations for the outer domain.

These approaches to the approximation of conditions at infinity are, of course, applicable to any of the methods described.

The "artificial stabilization method" provides the greatest "freedom" in the choice of computational procedures. Just as for systems of parabolic equations the latest methods of decomposition of computational operator can be applied in the most natural way. More precisely, these are the methods which reduce the solution of a algebraic linear system of high order to the solution of some set of systems of lower order.

For the system (VI) we used several methods; two of them deserve to be mentioned here.

Alternating directions method in which the space derivativatives in ξ and η directions are written in turn in implicit form.

The first time "half step":

$$\frac{\psi_{n+1} - \psi_n}{\Delta t} = \frac{\partial^2 \psi_{n+1}}{\partial \xi^2} + \frac{\partial^2 \psi_n}{\partial \eta^2} - F(\xi,\eta)\omega_n$$

$$\frac{\omega_{n+1} - \omega_n}{\Delta t} = \frac{\partial^2 \omega_{n+1}}{\partial \xi^2} + \frac{\partial^2 \omega_n}{\partial \eta^2} - 2 \left[\frac{\partial \psi_{n+1}}{\partial \eta} \cdot \frac{\partial \omega_{n+1}}{\partial \xi} - \frac{\partial \psi_{n+1}}{\partial \xi} \cdot \frac{\partial \omega_n}{\partial \eta} \right]$$

$$\frac{\omega_{n+1} - \omega_n}{\Delta t} = \alpha(\xi, t_{n+1}) \frac{\partial \psi_{n+1}}{\partial \eta} \qquad (\eta = \text{const})$$

The second time "half step":

$$\frac{\psi_{n+2} - \psi_{n+1}}{\Delta t} = \frac{\partial^2 \psi_{n+1}}{\partial \xi^2} + \frac{\partial^2 \psi_{n+2}}{\partial \eta^2} - F(\xi,\eta)\omega_{n+1}$$

$$\frac{\omega_{n+2} - \omega_{n+1}}{\Delta t} = \frac{\partial^2 \omega_{n+1}}{\partial \xi^2} + \frac{\partial^2 \omega_{n+2}}{\partial \eta^2} - 2 \left[\frac{\partial \psi_{n+2}}{\partial \eta} \cdot \frac{\partial \omega_{n+1}}{\partial \xi} - \frac{\partial \psi_{n+2}}{\partial \xi} \cdot \frac{\partial \omega_{n+2}}{\partial \eta} \right]$$

$$\frac{\omega_{n+2} - \omega_{n+1}}{\Delta t} = \alpha(\xi, t_{n+2}) \frac{\partial \psi_{n+2}}{\partial \eta} \qquad (\eta = \text{const})$$

For simplicity we write here the space derivatives in differential form. In calculations they are, of course, replaced by corresponding finite-difference expressions. It is important to note that the three-point approximation is used, for instance,

$$\frac{\partial^2 \omega_n}{\partial \xi^2} \cong \frac{\omega_n^{k+1} - 2\omega_n^k + \omega_n^{k-1}}{\Delta \xi^2}$$

Just this three-point scheme reduces the problem to the solution of a three-diagonal system of algebraic linear equations.

Locally onedimensional method. Here the following sequence of systems is solved:

The first half-step

$$\frac{\psi_{n+1} - \psi_n}{\Delta t} = 2 \frac{\partial^2 \psi_{n+1}}{\partial \xi^2}$$

$$\frac{\omega_{n+1} - \omega_n}{\Delta t} = 2 \frac{\partial^2 \omega_{n+1}}{\partial \xi^2} - 4 \frac{\partial \psi_{n+1}}{\partial \eta} \cdot \frac{\partial \omega_{n+1}}{\partial \xi}$$

The second half-step

$$\frac{\psi_{n+2} - \psi_{n+1}}{\Delta t} = 2 \frac{\partial^2 \psi_{n+2}}{\partial \eta^2} - 2F(\xi,\eta)\omega_{n+1}$$

$$\frac{\omega_{n+2} - \omega_{n+1}}{\Delta t} = 2 \frac{\partial^2 \omega_{n+2}}{\partial \eta^2} + 4 \frac{\partial \psi_{n+2}}{\partial \xi} \cdot \frac{\partial \omega_{n+2}}{\partial \eta}$$

The boundary conditions are approximated as in previous cases.

We see in this method there is no approximation of differential equations at each half-step. But two half-steps together give

$$\frac{\psi_{n+2} - \psi_n}{2 \Delta t} = \frac{\partial^2 \psi_{n+1}}{\partial \xi^2} + \frac{\partial^2 \psi_{n+2}}{\partial \eta^2} - F(\xi,\eta)\omega_{n+1}$$

$$\frac{\omega_{n+2} - \omega_n}{2 \Delta t} = \frac{\partial^2 \omega_{n+1}}{\partial \xi^2} + \frac{\partial^2 \omega_{n+2}}{\partial \eta^2} - 2 \left[\frac{\partial \psi_{n+1}}{\partial \eta} \cdot \frac{\partial \omega_{n+1}}{\partial \xi} - \frac{\partial \psi_{n+2}}{\partial \xi} \cdot \frac{\partial \omega_{n+2}}{\partial \eta} \right]$$

which is the approximation of the differential system.

To give "equal weight" to both variables ξ and η the alternating of them is used at successive full steps, let us say: first full step - at first the derivatives with respect to ξ remain in the equations and at the second half step the derivatives with respect to η remain; second full step - first half-step, derivatives in η remain, second half-step, derivatives in ξ remain and so on: ξ, η; ξ, η; ξ, η....

In the real stabilization method the procedures described can be applied only to the equation for the vorticity. For the stream-function some other method must be used, which gives for each time step the solution of the elliptical Poisson equation, by the matrix-factorization method, for instance.

This method is not good enough for large Reynolds numbers, because it requires a large computer memory. Therefore some other methods must be used (for instance, based on Fourier expansions) which permit some economy of memory, although requiring many more arithmetical operations.

Finally I shall mention one important computational procedure which permits us to simplify calculations in the case of a flow domain of complicated shape.

The simple shapes of domain (for example - rectangles) permit us to reduce the amount of calculation and the logic of the computer program also. If some complicated domain can be subdivided in a set of simple sub-domains, it is reasonable to carry out the solution for each sub-domain connecting them on the boundaries. For illustration of the idea let us consider the flow in Borda's mouthpiece (Fig. 1). We construct the solution of the Navier-Stokes system in sub-domains 1 and 2 (SD1 and SD2 on Fig. 1). On the common border (dotted line) the following conditions must be fulfilled:

$$\frac{\partial \omega_1}{\partial \eta} = \frac{\partial \omega_2}{\partial \eta} \quad ; \quad \omega_1 = \omega_2 \quad ; \quad \frac{\partial \psi_1}{\partial \eta} = \frac{\partial \psi_2}{\partial \eta} \quad ; \quad \psi_1 = \psi_2$$

Instead of this system of four conditions we write two conditions for each sub-domain:

SD1

$$\frac{\partial \psi_1}{\partial \eta} = \beta(\psi_1 - \psi_2) + \frac{\partial \psi_1}{\partial \eta}$$

$$\frac{\partial \omega_1}{\partial \eta} = \gamma(\omega_1 - \omega_2) + \frac{\partial \omega_1}{\partial \eta}$$

SD2

$$\frac{\partial \psi_2}{\partial \eta} = \beta(\psi_1 - \psi_2) + \frac{\partial \psi_2}{\partial \eta}$$

$$\frac{\partial \omega_2}{\partial \eta} = \gamma(\omega_1 - \omega_2) \quad \frac{\partial \omega_2}{\partial \eta}$$

Using successive approximations we have

$$\frac{\partial \psi_{1,n+1}}{\partial \eta} = \beta(\psi_{1,n} - \psi_{2,n}) + \frac{\partial \psi_{1,n}}{\partial \eta}$$

$$\frac{\partial \omega_{1,n+1}}{\partial \eta} = \gamma(\omega_{1,n} - \omega_{2,n}) + \frac{\partial \omega_{1,n}}{\partial \eta}$$

and similarly for the second sub-domain. If the initial approximation ensures that

$$\frac{\partial \psi_{1,0}}{\partial \eta} \equiv \frac{\partial \psi_{2,0}}{\partial \eta} \quad , \quad \frac{\partial \omega_{1,0}}{\partial \eta} \equiv \frac{\partial \omega_{2,0}}{\partial \eta}$$

then the equality between $\partial \omega_1/\partial \eta$ and $\partial \omega_2/\partial \eta$ will always be fulfilled (the same for $\partial \psi_1/\partial \eta$ and $\partial \psi_2/\partial \eta$). If the successive approximations converge then automatically the condition $\omega_1 = \omega_2$, $\psi_1 = \psi_2$ will be satisfied. We see that at each step of the iteration the equations for each sub-domain are solved separately.

The convergence of the method (by proper choice of the relaxation parameters β or γ) is easily proved for Laplace's equation. For the Navier-Stokes system the convergence was verified only by calculations.

Quite naturally the question can arise: which method is better?

I answer quite definitely - the stationary method of successive approximation (if convergent, of course). Its most satisfying property is that the rate of convergence does not depend on the accuracy of the approximation (upon the number of nodal points in finite-difference grid). In any stabilization method the time-step decreases when the finite-difference grid becomes denser.

Nevertheless the application of different methods gives some possibility for better understanding of the processes in viscous fluids. I will make some further remarks on this.

III. Some Results of Calculation

Now the number of different cases calculated is already so huge all over the world that there is no reason to show many "picturesque images". I limit myself therefore to a few results which I selected, considering them to be "suggestive" in some respect.

As the first example consider the motion in an expanding channel (Fig. 2). The flow picture here has no surprises.

At small Reynolds numbers separation does not occur. When the Reynolds number increases the separation appears, the point of detachment very soon stabilizes, while the reattachment point moves off roughly proportionally to Reynolds number (Fig. 3).

I give this example because it was a test case in our calculations for which different methods were tried out. In Fig. 4, for example, the comparison of calculations by the method of successive approximation and by artificial stabilization method (using locally one-dimensional procedure) is represented. Here Re - 100 (32 π).

This use of different methods disclosed one interesting phenomenon: all the methods fail when Re is close to 200.

The properties of the methods are very different, and the easiest way to explain their simultaneous failure is to suppose that the solution of the steady flow equations does not exist for such large Reynolds numbers.

Of course, numerical methods can never give the proof of some mathematical fact, but they are suggestive at least.

In connection with this we intend to undertake wide numerical experiments with the use of different methods for the same cases of flow with separation. All the calculations for this case were made by Miss N. Meller - the senior scientific collaborator of the Computing Center.

Figures 4 and 5 show the flow around fixed and rotating cylinders (the example is taken from the work of Dr. V. Lulka).

The separation vanishes when the rotation speed is large enough.

The third example illustrates the motion in a "pit" - in the channel with local expansion. Here the method of subdividing of domain was used (dotted lines on Fig. 6). The results of the calculation were kindly put at my disposal by the Bulgarian postgraduate Miss E. Mateyeva.

The picture of flow itself is interesting here. It shows how careful one must be when trying to use the integral conservation laws (mass, impulse, energy) on the basis of some flow picture suggested a priori.

I am not showing here any examples of compressible flow calculations, for although they include some interesting results, we are still in the initial stage of investigating methods in this case.

IV. The Problem of Large Reynolds Numbers

In assessing the possibility of the numerical solution of the Navier-Stokes equations in the present state of computer capacity, we can say that this is realizable for an incompressible fluid in plane or axisymmetrical flow, and perhaps even easy for practical calculations when Reynolds numbers are not too large (of the order one hundred, approximately). Unfortunately, the most interesting problems in practice relate to much larger Reynolds numbers.

The difficulties here, I would say, are purely computational. In a finite - difference approximation to the nonlinear term of equations the truncation error is in general of the order $Re \cdot h$ ($h = \Delta x$ or Δy), when the order of main viscous term is supposed to be of the order 1. Obviously, an approach to the real solution can be reached when $Re \cdot h < \varepsilon \ll 1$, otherwise the mathematical viscosity exceeds the real one and we don't know what the numerical solution means.

It seems that by using central-difference expressions for derivatives we can reduce the error to the order $Re \cdot h^2$ and the condition of approximation will be $Re \cdot h^2 < \varepsilon \ll 1$, which is much more favorable than the previous one. However, the truncation error in this case contains the third derivatives of vorticity, and we can't be sure that ε is independent of Re. To imagine what it means, let us evaluate, for example, the work for solution of an unsteady problem in the case of Poiseuille motion between two plates. According to C. C. Lin's calculation the breakdown of stability of Poiseuille motion occurs when $Re > 6000$. Bearing in mind that we wish to obtain the picture of development of turbulence by numerical calculations, we need to satisfy the condition, let us say, $Re \cdot h^2 = 0.1$, that is, $h \cong 1/250$. The length of channel must be at least 20 times more than the width. So our finite - difference grid must have at least 5000 nodal points.

Even with all the achievements of modern computer technology this is a horrible problem. And it must be noted that the main obstacle now is the volume of fast memory - not the speed of computers. Even with the speeds which are already attainable in a good computer, the problem would have been attacked, if the fast memory contained something like 10^7 words.

These estimates, unfavorable as they are for the time being, nevertheless show a good prospect for the future (and even the near future). Computers will soon have such a memory, together with an increase of speed.

In conclusion I would like to make a remark. For practical application of numerical solutions of the Navier-Stokes equations it is not always necessary to apply them over the whole domain of flow. We have to learn to combine the solutions of this system in different approximations: boundary layer, ideal fluid approximations. To use "dogmatically" the complete Navier-Stokes equations for the whole field of flow is not only difficult, but even unreasonable: the accuracy of results can deteriorate instead of increasing.

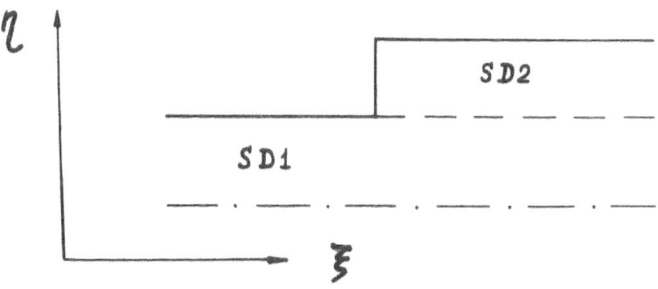

Fig. 1

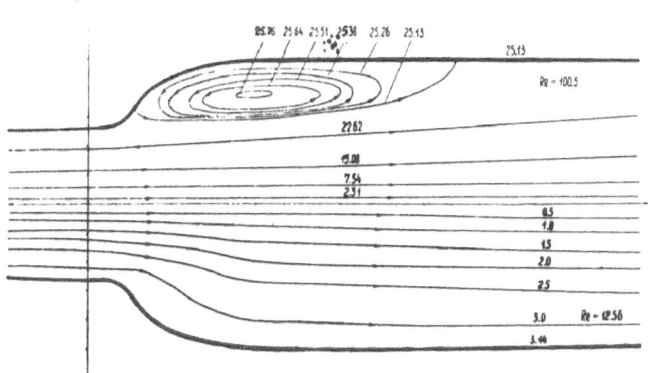

Fig. 2

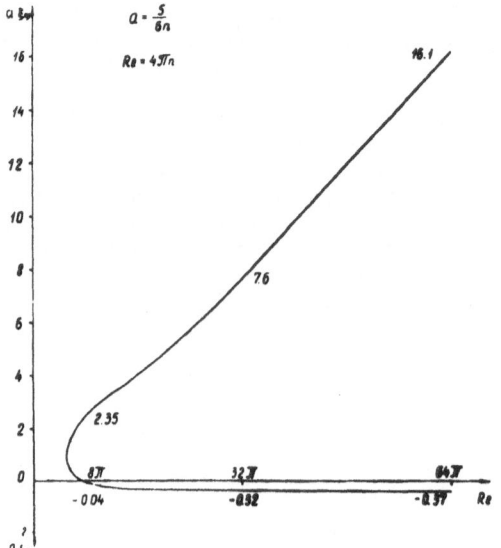

Fig. 3

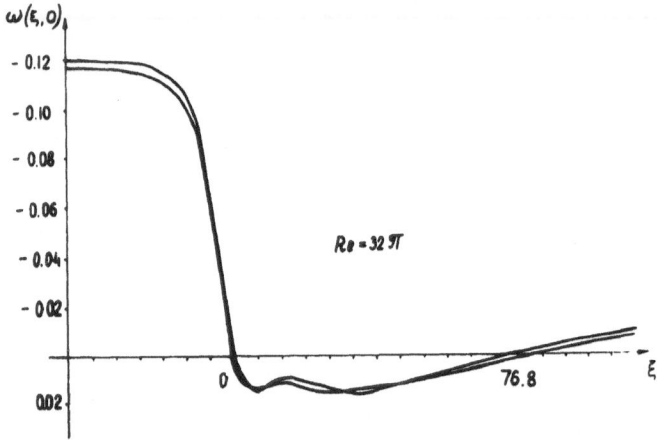

Fig. 4

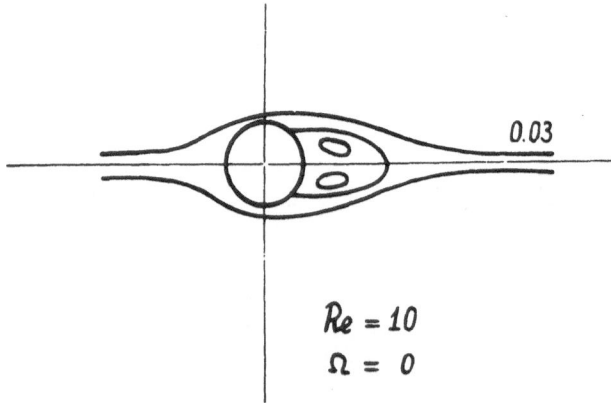

Fig. 5

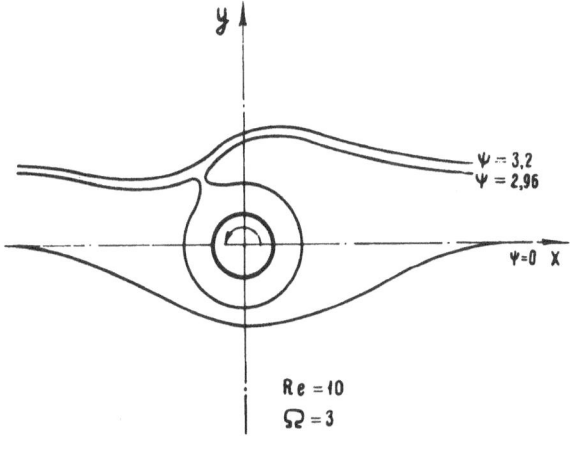

Fig. 6

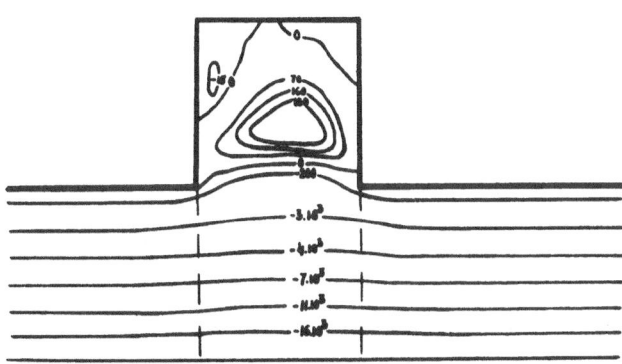

Fig. 7

ATMOSPHERIC DYNAMICS AND THE NUMERICAL

SIMULATION OF ATMOSPHERIC CIRCULATION

by Pierre MOREL
Professor, University of Paris VI

1. INTRODUCTION

Since the advent of automatic computing techniques, many
attempts have been made to forecast the meteorologically significant
components of atmospheric motions using various statistical or
dynamical hypotheses to simplify matters and a fair amount of success
has been met in actual operational weather prediction with these
methods. We shall not address ourselves to this practical problem
here, but consider rather the more fundamental question of actually
simulating the dynamics of atmospheric motions with the help of a
numerical analogue obeying the same dynamical and thermodynamical
laws, i.e. the Navier-Stokes equations in their essentially primitive
form. The first successful numerical integration of the primitive
equations of fluid dynamics applied to atmospheric circulation is
due to PHILLIPS (1956). Since this day, many groups in the world
have refined the numerical integration techniques and advanced our
knowledge of the various energy sources and sinks in the atmosphere
to the extent that it can be said that reasonnably realistic
numerical analogues or models of the general circulation of the
atmosphere are feasible and are indeed operated by several groups;
see: SMAGORINSKY, et al. (1965); MANABE et al. (1965); KASAHARA and
WASHINGTON (1967, 1971); SHUMAN and HOVERMALE (1968) for accounts
of the first achievements in this endeavour. Similar techniques
could certainly be applied to the planetary scale circulation in

in large oceanic basins : due to the added complication of exceedingly variable bottom topography and far lesser knowledge of significant
scales in the sea, oceanic circulation models have not progressed as
far as atmospheric models; see for example: BRYAN and COX (1967),
HOLLAND (1967), CROWLEY (1968) and BRYAN (1969). The numerical
simulation of other smaller scale flows of geophysical importance
like hurricanes, thermal convection, three-dimensional turbulent
flow in the planetary boundary layer are rather less advanced,
although promising results have been achieved; see for example:
OOYAMA (1969), SUNDQUIST (1970) and ROSENTHAL (1970 a,b) on hurricanes;
LILLY (1962), MURRAY (1970) on cumulus convection; DEARDORFF (1970
a,b) on the turbulent boundary layer. Finally, atmospheric
physicists have already made the first step toward simulating other
planets' atmospheric circulation with appropriately intriguing (and
controversial) results: LEOVY and MINTZ (1969); SASAMORI (1971). In
view of the definitely more advanced stage reached by numerical
modelling of planetary scale geophysical flows like the general
circulation of the Earth atmosphere - and also because this is the
author's personal field of interest - we shall henceforth restrict
our attention to this subject. In the first part, we shall survey
the special features of large scale planetary flow which make the
numerical simulation of atmospheric circulation a distinctly original
problem, quite separate from standard hydro or aerodynamical problems,
say. In the second part, we shall take the first steps toward
introducing the numerical schemes applicable to the integration of
the primitive partial differential equations of fluid dynamics.
Finally, in the last part, we will give an overview of the frontier
yet to be conquered, outstanding questions to be solved before one
could successfully tackle the problem of predicting not only the next
48 or 72 hours weather but the possible future climates of our planet
in response to different hypotheses regarding the action of man on
its natural environment.

PART ONE

2. THE STATIC ATMOSPHERE

The zero-order approximation to the theory of planetary atmospheres is a static, horizontally homogeneous fluid, vertically stratified according to a standard density profile

$$\rho_s = \rho_s(z) \tag{2.1}$$

as shown in Fig.1. Also shown is a discrete approximation of this density profile by a number of successive layers of finite thicknesses as one would want to use for a finite difference approximation of the governing equations. The pressure everywhere in the fluid derives from (2.1) and the hydrostatic relation:

$$\frac{\partial p}{\partial z} = - \rho g \tag{2.2}$$

It is also a standard function $p_s(z)$ of altitude z only. Finally, the temperature $T_s(z)$ follows from the equation of state:

$$\frac{p}{\rho} = RT \tag{2.3}$$

The density of the atmospheric fluid decreases very rapidly (quasi exponentially) with altitude but never vanishes completely so that the Earth's upper atmosphere can be said to reach extreme distances like several Earth radii from the planetary surface. For practical meteorological purposes, however, the Earth atmosphere is quite thin since 90% of its mass lies below 16 km and 99% below 30 km. One order of magnitude of the atmosphere depth is the scale height H above which the density is reduced by a factor 1/e:

$$H = \frac{p_s}{g\rho_s} = \frac{RT_s}{g} \simeq 8 \text{ km} \tag{2.4}$$

15

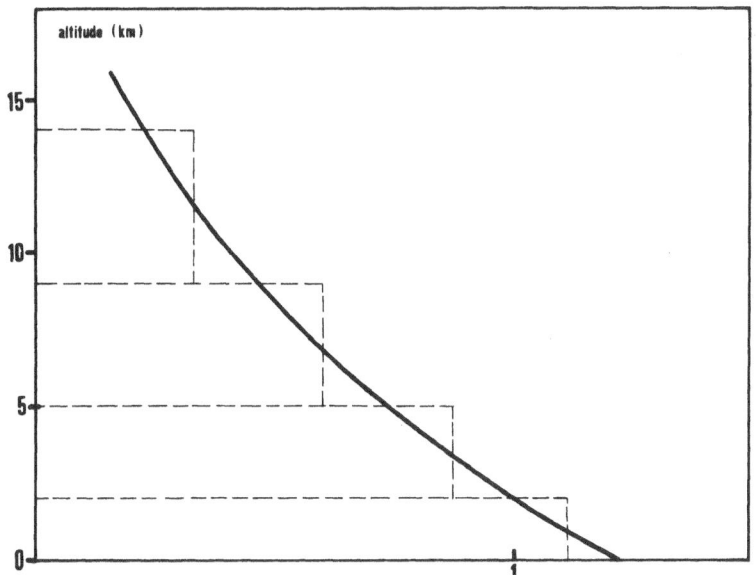

Figure 1
Density the standard atmosphere

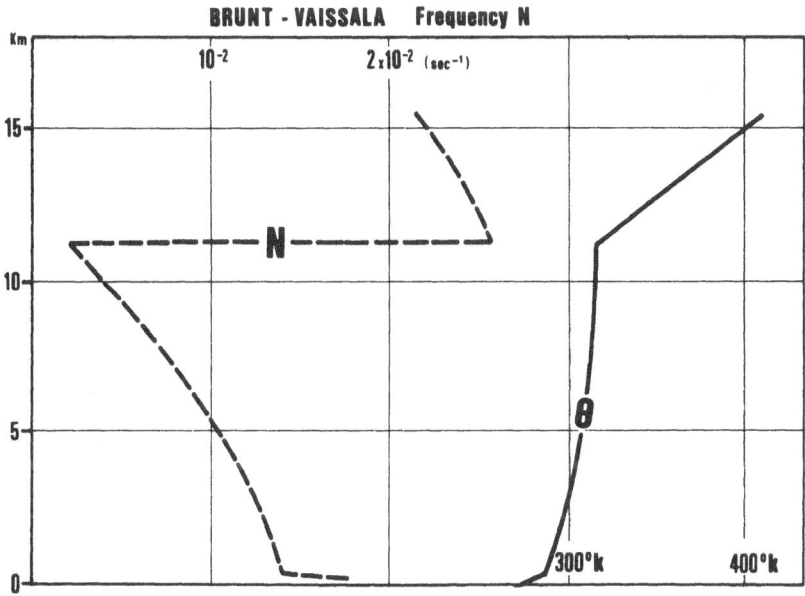

Figure 2
Potential temperature θ

Most important is the fact that the planetary atmosphere is under
(generally) <u>stable vertical stratification</u>. Stable or unstable
stratification depends upon the increase or decrease of the fluid
entropy S with height:

$$S = C_p \log T - R \log p + c^{te}$$

or rather:

$$S = C_p \log \theta + c^{te} \qquad\qquad (2.5)$$

introducing the potential temperature $\theta = T(p/p_o)^{-R/C_p}$. Now, we
can see that the potential temperature (entropy) of a quickly
rising (or subsiding) air parcel will not change because the
corresponding expansion (compression) will be essentially adiabatic.
If the standard potential temperature profile $\theta_s(z)$ is increasing
with height, a rising air parcel becomes colder and heavier than
the surrounding atmosphere, and comes down again, thus beginning
a vertical oscillation known as the BRUNT-VAÏSSALA oscillation .
The BRUNT-VAÏSSALA frequency N is given by:

$$N^2 = g \frac{\partial \log \theta_s}{\partial z} \qquad\qquad (2.6)$$

where N^2 positive means a stable stratification and N^2 negative
an unstable stratification. Although unstable conditions do occur
intermittently in the atmosphere, they are quickly relaxed by
cellular convection (giving rise to cumulus clouds and a large
release of latent heat): they may therefore be ignored in large
scale planetary flows. Figure 2 is a typical mid-latitude potential
temperature profile and the associated BRUNT-VAÏSSALA
frequency profile; both plots show two distinctly different regions,
to wit, the troposphere characterized by moderately stable or even
neutral stratification and the stratosphere considerably more stable.
This of course has a drastic effect on atmospheric dynamics; the
stratosphere is so stable that this upper part of the atmosphere
acts almost like an inert lid where no conversion of the internal
energy into mechanical energy can take place. For this reason,

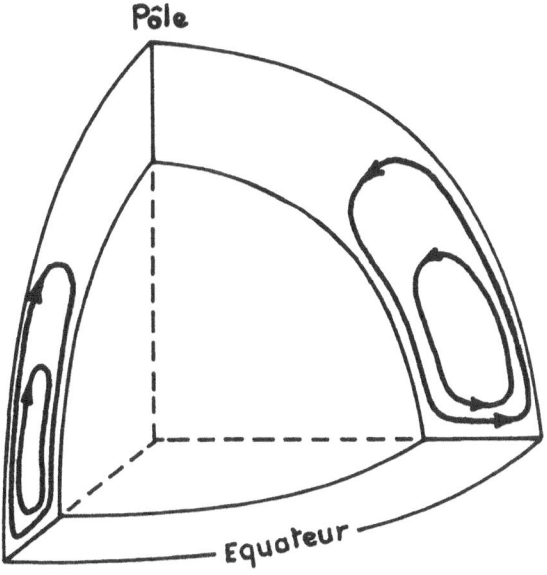

Figure 3
The mean meridional (HADLEY) circulation in
the tropics

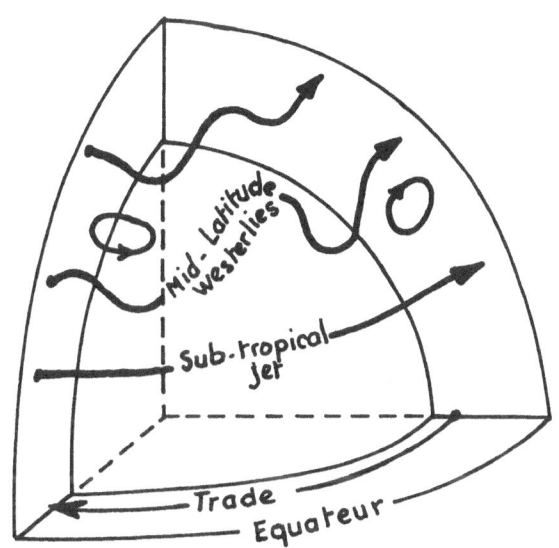

Figure 4
The mean and eddy zonal circulation

we may, to a first order of approximation, restrict our attention
to the troposphere where most of the action takes place, regarding
the production of the mechanical energy which drives the atmosphere
and oceans. This region extends from the surface up to the
tropopause; its thickness D varies from 8 km in polar regions to
18 km or so in the Tropics. For the purpose of scale analysis
we may typically take:

$$D \simeq 10 \text{ km}$$

The troposphere is thus characterized by a rather weakly stable
stratification:

$$D \frac{\partial \log\theta_s}{\partial z} \simeq 0.1 \text{ to } 0.2 \qquad (2.7)$$

3. THE MEAN ZONAL CIRCULATION

If the atmosphere were horizontally homogeneous everywhere,
the thermodynamically stable state of motion would be solid rotation
at the same angular velocity Ω as the planetary core just like the
solid rotation of a (viscous) fluid in a spinning cylindrical
container. But the Earth atmosphere is not horizontally homogeneous
because, for a given pressure $p_s(z)$, the air is colder and denser
near the Poles than near the Equator. This is just the opposite
of a stable situation because equilibrium in a rotating fluid
system requires that the denser parts be farther away from the
rotation axis. The temperature (and density) contrast prevailing
on the Earth and similar radiatively heated planets, sets a meridional circulation of the cold polar air toward the Equator and a
corresponding counter-current above (Figure 3). Although this
circulation tends to restore equilibrium conditions, this state of
equilibrium is never actually reached because of continuous
radiative heating of the ground and lower layer of the atmosphere

on the one hand, and continuous radiative cooling of the upper
layer on the other hand. Thus the troposphere behaves like a
continuously cycling thermal engine with a warm heat source (the
ground or sea surface) which receives its energy from the sun, and
a cold heat sink (deep space).

Now, it is easy to see that under these circumstances,
conservation of angular momentum produces easterly wind near the
ground in the Tropics and a strong upper altitude westerly flow in
the mid-latitude. These currents are indeed observed in the zone
between 30 North and 30 South where this coupled meridional-zonal
circulation (HADLEY circulation) is quite stable and produces the
well known easterly trade winds at low altitude and a very strong
(100 knots) westerly sub-tropical jet in the upper atmosphere.

It is of interest to estimate the order of magnitude of the
kinetic energy $\overline{K}$ of the general circulation of the atmosphere,
relative to whatever energy would be stored anyway in a static
atmosphere. Now, the static atmosphere has both gravitational
potential energy:

$$\overline{P} = \int_0^\infty \rho g z \, dz = \int_0^{p_o} RT \, \frac{dp}{g} \qquad (3.1)$$

and internal (thermal) energy:

$$\overline{I} = \int_0^{p_o} C_v \, T \, \frac{dp}{g} \qquad (3.2)$$

Here dp/g is the mass in a column of height dz and unit cross
section. Relation 3.1 is arrived at by use of the hydrostatic
equation (2.2). Now (3.1) and (3.2) can be combined to yield the
total potential energy of the static atmosphere:

$$\overline{E} = \overline{P} + \overline{I} = \int_0^{p_o} C_p T \, \frac{dp}{g} = \frac{C_v}{R} \int_0^{p_o} c^2 \, \frac{dp}{g} \qquad (3.3)$$

where $C^2 = RT\, C_p/C_v$ is the square of the velocity of sound in air. On the other hand, the total kinetic energy in a column of unit cross section is:

$$\bar{K} = \int_0^{p_0} \frac{1}{2}\, V^2\, \frac{dp}{g} \qquad (3.4)$$

So that:

$$\frac{\bar{K}}{E} \simeq \frac{1}{2}\, \frac{R}{C_v}\, \frac{\bar{V}^2}{\bar{C}^2} \simeq \frac{1}{5}\, \frac{\bar{V}^2}{\bar{C}^2}$$

where we have substituted $C_v = \frac{5}{2}\, R$; $\bar{V}^2$ and $\bar{C}^2$ are appropriate mean values of the atmospheric circulation velocity (about 15 m sec^{-1} say) and the speed of sound (about 300 m sec^{-1}) respectively. Thus:

$$\frac{\bar{K}}{E} \simeq \frac{1}{2000}$$

Now it is true that the total potential energy of the atmosphere is not <u>available</u> to be converted into kinetic energy: only that part of the potential energy which corresponds to departures from the standard horizontal stratification can readily generate kinetic energy as shown by LORENZ (1955). The available potential energy is only about ten times as large as $\bar{K}$. The fact remains nevertheless that the circulation of the atmosphere is a very small perturbation indeed of the static atmosphere and it is correspondingly <u>difficult to compute</u>, for even the smallest systematic error on the distribution of mass would cause drastic deformations of the wind patterns. We shall remember that any computational scheme to integrate the equations of motion of geophysical flow must satisfy exact detailed balancing of inbound and outbound mass fluxes or equivalently exact conservation of the total mass of the fluid.

4.   SCALE ANALYSIS FOR THE PLANETARY

CIRCULATION OF THE ATMOSPHERE

Before we may proceed, we must now introduce the dynamic (Navier-Stokes) and thermodynamic equations applicable to the general circulation of the atmosphere on a rotating planet and determine the relative order of magnitude of the different terms which appear in these equations.

Let us begin by introducing a typical horizontal scale L over which the velocity of the planetary flow changes appreciably, i.e. about 1000 km for the main perturbations or waves which contain a larger part of the eddy kinetic energy of the atmospheric circulation. A very significant dimensionless parameter is then the ratio of the typical relative velocity U of the flow, to the variation of the solid rotation velocity of the planet over the distance L:

$$R_o = \frac{U}{2\Omega L} \qquad\qquad (4.1)$$

Clearly this number (the ROSSBY number) serves to distinguish the circulation on slowly rotating planets ($R_o \gtrsim 1$) like Venus from the circulation on <u>rapidly rotating</u> planets ($R_o \ll 1$) like Mars, Jupiter and, of course, the Earth. For Earth $\Omega = 7.3 \times 10^{-5}$ rad.sec^{-1} and $R_o \simeq 0.1$ typically.

Next, we may take for vertical scale D the thickness of the troposphere or a large fraction of it. It is immediately evident that our atmosphere is <u>thin</u> as:

$$\frac{D}{L} \simeq 0.01 \quad to \quad 0.001 \qquad\qquad (4.2)$$

is much smaller than 1. Consequently, we expect vertical velocities w in the flow to be much smaller than horizontal velocities, to wit:

$$w \lesssim \frac{D}{L} U \qquad\qquad (4.3)$$

There are in fact fairly small scale motions of the atmospheric fluid
which produce vertical velocities of the order of UD/L. We shall
see however, that these motions do not lend themselves well to
converting thermal into mechanical energy and are not important for
understanding or simulating the atmosphere dynamics. These motions
are gravity waves, both internal and external, quite similar to
waves on the sea surface.

Now, motions which do convert the available potential energy
of the atmosphere into mechanical energy tend to equalize the
vertical stratification while increasing the gravitational potential
energy. In such a motion, an air parcel will therefore follow a
trajectory intermediate between the slope of isentropic surfaces
(θ = constant), and equipotential surfaces (z = constant). Thus:

$$\frac{w}{U} \; \lesssim \left(\frac{\partial z}{\partial y}\right)_{\theta \; = \; constant}$$

where y is taken in the horizontal direction. It can be seen that:

$$\left(\frac{\partial z}{\partial y}\right)_{\theta = c^t} \simeq \; -\left(\frac{\partial \log \theta_s}{\partial y}\right)\left(\frac{\partial \log \theta_s}{\partial z}\right)^{-1}$$

$$\simeq \; Ro \; \frac{D}{L}\left(\frac{4\Omega^2 L^2}{gD}\right)\left(D\frac{\partial \log \theta_s}{\partial z}\right)^{-1} \qquad (4.4)$$

The factor $4\Omega^2 L^2/gD$ is a dimensionless parameter (FROUDE number Fr)
i.e. the ratio of the kinetic energy of the solid rotation of the
planet (or a typical variation of the same over the distance L) to
the gravitational potential energy at the top of the troposphere.
For Earth, the FROUDE number is of the order of 0.2. We have seen
above (2.7) that the second factor between parentheses is also a
dimensionless number of the order of 0.1 or 0.2. We may conclude
then:

$$w \simeq \; Ro \; \frac{D}{L} \; U \qquad (4.5)$$

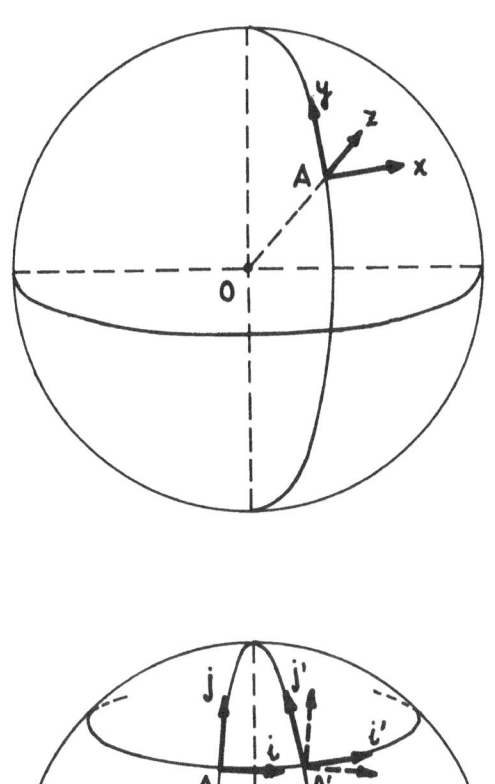

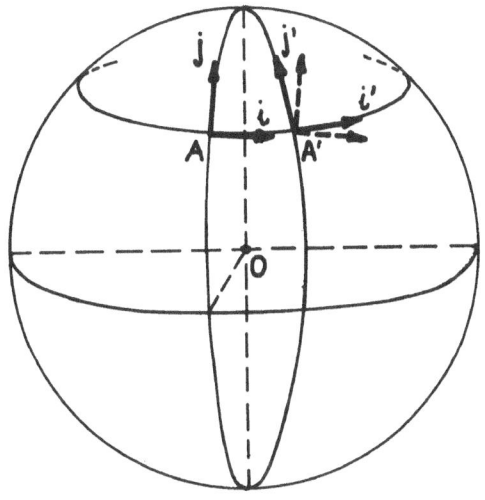

Figure 5
Local coordinate system and
sphericity correction (rotation)

for the energetically significant motions of planetary scale geophysical flows. Typically then, the vertical velocity associated
with these large scale motions may be of the order of 10^{-2} m sec^{-1}
and very much smaller than the typical horizontal velocity. Further,
the pressure differences $\Delta p = p - p_s(z)$ arrising from the large
scale, quasi-horizontal components of the flow, can be estimated
from the (approximate) balance between the horizontal pressure
gradient and the major horizontal acceleration of the fluid which
is the Coriolis acceleration. Thus:

$$\frac{1}{\rho_s} \frac{\Delta p}{L} \simeq 2\Omega U$$

or:

$$\frac{\Delta p}{p_s} = \frac{2\Omega L U \rho_s}{p_s} = Ro \; Fr \; \frac{D}{H} \tag{4.6}$$

The corresponding order of magnitude of the density and potential
temperature (entropy) differences follow:

$$\frac{\Delta \rho}{\rho_s} \simeq Ro \; Fr \tag{4.7}$$

$$\frac{\Delta \theta}{\theta_s} \simeq Ro \; Fr \tag{4.8}$$

It can be seen from (4.6)-(4.8) that the typical mass and temperature
distributions in the actual atmosphere differ by a very small amount,
1% or so, from the standard stratification of the static atmosphere.
Yet these rather small pressure and temperature contrasts can
produce quite a rapid atmospheric circulation.

Let us now review the governing equations of the planetary
atmosphere circulation in the light of this brief analysis of
orders of magnitude. We introduce local coordinates x,y and z
as shown in Fig. 5, unit vectors $\vec{i}$, $\vec{j}$ and $\vec{k}$ along the directions
Ax, Ay and Az respectively, and the three components u,v and w of
the flow velocity. We shall, for simplicity, neglect sphericity
corrections arrising from using local coordinates on a curved

surface (the Earth spheroid) : these corrections are in fact taken into account in a complete treatment of the dynamic problem. The Navier-Stokes equations for the horizontal velocity $\vec{V}(u,v)$ and vertical velocity w on a rotating planet read:

$$\left(\frac{\partial}{\partial t} + \vec{V}.\nabla + w\frac{\partial}{\partial z}\right)\vec{V} + 2\Omega\,\sin\phi\,\vec{k}\times\vec{V} + 2\Omega\,\cos\phi\,w\,\vec{i} + \frac{1}{\rho}\,\nabla p$$

$$= \begin{cases} 0 \text{ if no viscous stress} \\[2mm] \nu\nabla^2\,\vec{V} \text{ otherwise} \end{cases} \qquad (4.9)$$

$$\left(\frac{\partial}{\partial t} + \vec{V}.\vec{\nabla} + w\,\frac{\partial}{\partial z}\right)w - 2\Omega\,\cos\phi\,\vec{V}.\vec{i} + g + \frac{1}{\rho}\,\frac{\partial p}{\partial z} = 0 \qquad (4.10)$$

where ϕ is the latitude and Ω the angular velocity of the planet. We have the thermodynamic equation:

$$\left(\frac{\partial}{\partial t} + \vec{V}.\nabla + w\,\frac{\partial}{\partial z}\right)\log\theta = \begin{cases} 0 \text{ if adiabatic} \\[2mm] \dfrac{\dot{Q}}{C_p T} \quad \text{otherwise} \end{cases} \qquad (4.11)$$

the continuity equation:

$$\frac{\partial\rho}{\partial t} + \nabla\cdot(\rho\vec{V}) + \frac{\partial}{\partial z}\,(\rho w) = 0 \qquad (4.12)$$

and finally the equation of state (2.3).

We may now scale all quantities according to the orders of magnitude found above and expand the solution in series of the (small) dimensionless ROSSBY number Ro. The <u>zero order</u> solution is found to satisfy:

$$\left.\begin{aligned} f\vec{k}\times\vec{V} + \frac{1}{\rho_s}\,\nabla p &= 0 \\[3mm] \nabla\cdot(\rho_s\vec{V}) &= 0 \end{aligned}\right\} \qquad (4.13)$$

It is thereforeanon divergent, horizontal flow in geostrophic
balance where the pressure gradient force exactly balances the
Coriolis force $f\vec{k} \times \vec{V}$; f is the Coriolis parameter $2\Omega\sin\phi$.

The <u>first order</u> equations allow for vertical motion. The
standard density stratification of the fluid $\rho = \rho_s(z)$ is used
everywhere but in the hydrostatic equation,because small density
differences coupled with gravity are the driving force of the
atmospheric circulation. The first order equations correspond
therefore to the BOUSSINESQ approximation. They read:

$$\left.\begin{array}{l} (\dfrac{\partial}{\partial t} + \vec{V}.\vec{\nabla})\vec{V} + f\vec{k} \times \vec{V} + \dfrac{1}{\rho_s}\nabla p = 0 \\[4mm] g + \dfrac{1}{\rho}\dfrac{\partial p}{\partial z} = 0 \\[4mm] (\dfrac{\partial}{\partial t} + \vec{V}.\vec{\nabla})\log\theta + w \dfrac{\partial\log\theta_s}{\partial z} = 0 \\[4mm] \nabla.(\rho_s\vec{V}) + \dfrac{\partial}{\partial z}(\rho_s w) = 0 \end{array}\right\} \qquad (4.14)$$

The <u>second order</u> equations are only slightly more complicated
and are generally used for large scale geophysical flow simulation
as the "<u>primitive equations</u>" of geophysical fluid dynamics:

$$\left.\begin{array}{l} (\dfrac{\partial}{\partial t} + \vec{V}.\nabla + w \dfrac{\partial}{\partial z})\vec{V} + f\vec{k} \times \vec{V} + \dfrac{1}{\rho}\nabla p = \nu\nabla^2\vec{V} \\[4mm] g + \dfrac{1}{\rho}\dfrac{\partial p}{\partial z} = 0 \\[4mm] (\dfrac{\partial}{\partial t} + \vec{V}.\nabla + w \dfrac{\partial}{\partial z})\log\theta = \dfrac{\dot{Q}}{C_p T} \\[4mm] \dfrac{\partial\rho}{\partial t} + \nabla.(\rho\vec{V}) + \dfrac{\partial}{\partial z}(\rho w) = 0 \end{array}\right\} \qquad (4.15)$$

Note that even to second order of the ROSSBY number, the vertical
acceleration vanishes and the dynamic equation for vertical motion
reduces to hydrostatic equilibrium (dw/dt is of order $Ro^2(D/L)^2$ or
10^7 times smaller than g). Large scale geophysical flows are thus
quasi-two-dimensional and this is indeed the reason why their
numerical simulation has been possible with existing computing
facilities : the simulation of true tri-dimensional turbulent flows
is just becoming feasible with the most powerful computers
available today.

5.   CHARACTER OF THE SOLUTIONS

The primitive equations (4.15) allow horizontal convergence
or divergence of the fluid only if compensated by corresponding
vertical mass fluxes: pure compressions and expansions like sound
waves are therefore excluded. A first insight into the nature of
the remaining permissible solutions can be gained by considering
a vertically homogeneous (or _barotropic_) flow $\vec{V} = \vec{V}(x,y,t)$ of
an incompressible fluid ($\rho_s = 1$). Taking equations (4.14) for
simplicity and integrating from z = 0 (where w = 0) to the height
h of the free surface (where w = dh/dt), and introducing the geopotential Φ = gh at the free surface, one gets :

$$\left.\begin{array}{l} (\frac{\partial}{\partial t} + \vec{V}.\nabla)\ \vec{V} + f\vec{k} \times \vec{V} + \nabla\Phi = 0 \\[2em] (\frac{\partial}{\partial t} + \vec{V}.\nabla)\ \Phi + \Phi\,\nabla.\vec{V} = 0 \end{array}\right\} \qquad (5.1)$$

We may linearize these equations by considering small perturbations u,v, Φ of a basic channel flow with uniform velocity $\vec{V}_o$
and free surface height H_o. Thus :

$$\frac{\partial u}{\partial t} = - \vec{V}_o . \nabla u + fv - \frac{\partial \Phi}{\partial x}$$

$$\frac{\partial v}{\partial t} = - \vec{V}_o . \nabla v - fu - \frac{\partial \Phi}{\partial x} \qquad (5.2)$$

$$\frac{\partial \Phi}{\partial t} = - \vec{V}_o . \nabla \Phi - gH_o \left(\frac{\partial u}{\partial x} + \frac{\partial v}{\partial y}\right)$$

We have now a set of linear differential equations with plane wave solutions like $\hat{\Psi}(k) \exp i(\omega t - kx)$ where $\hat{\Psi}(k)$ is a three dimensional eigenvector and w is one of three eigenvalues:

$$\omega_j(k) = \vec{V}_o . \vec{k} + j \sqrt{gH_o k^2 + f^2} \qquad (5.3)$$

$$(j = 1, 0, -1)$$

The eigenvector corresponding to eigenvalue ω_o (j=0) is a frozen wave, moving with the basic flow at the (phase) velocity $\vec{V}_o$. It does not produce any convergence or divergence as it progresses through advection only. In a more complete (non-linear) theory, similar solutions are found with phase velocities close to $\vec{V}_o$: they are quasi-geostrophic perturbations of the basic flow, producing only very small vertical velocities:

$$w \lesssim Ro \frac{D}{L} U$$

The j = 0 modes thus correspond to the meteorologically significant motions of the general planetary circulation, which convert available potential energy to mechanical energy.

The two other eigenvectors on the other hand (j=1 or -1) are waves with a much larger phase velocity of the order of C_o:

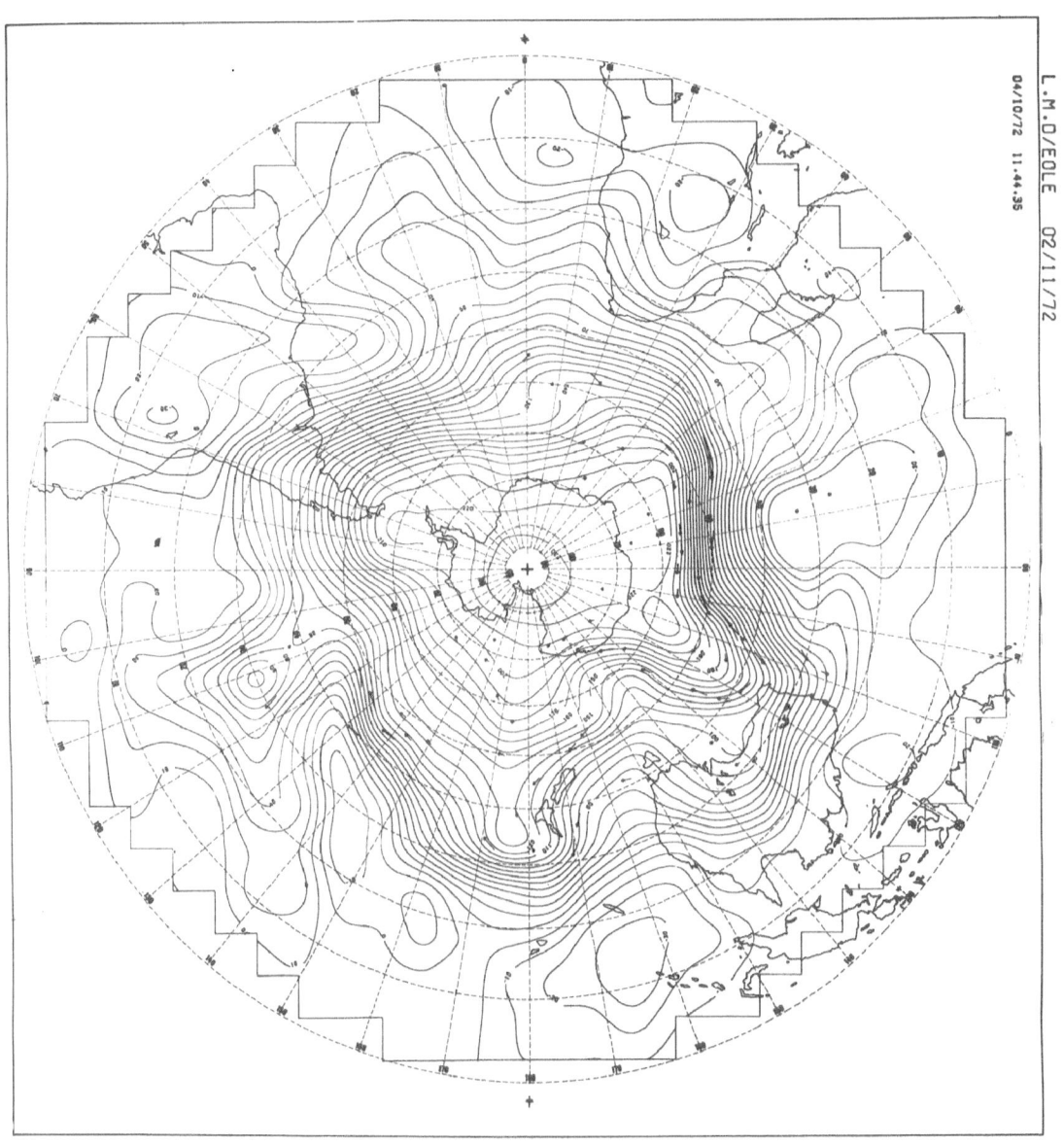

Figure 6
Streamfunction of the general circulation
at 200 mb in the Southern Hemisphere
from the EOLE experiment

$$\omega_1 \simeq \sqrt{gH_o k^2} = C_o k$$

where C_o is the velocity of external gravity waves, i.e. about 300 m sec^{-1} for the troposphere. The existence of such rapidly moving wavelike solutions of the primitive equations (4.15) is very significant for the numerical integration of these equations since the COURANT-LEOVY-FRIEDRICH stability condition requires that the fastest moving waves move no more than one resolution element or so during one time step Δt. These gravity waves are associated with relatively large horizontal divergence (convergence) and large vertical velocities $w \simeq UD/L$.

Now, very fast laminar flows cannot persist in a weakly stratified fluid because instabilities develop when a certain critical vertical wind shear is reached and large quasi-horizontal perturbations (belonging to modes j=0) are generated. The theories of this baroclinic instability (see for example MILES, 1964) indicate that the critical wind shear U/D in mid-latitudes is of the order of:

$$\frac{U}{D} \simeq \frac{N^2 D}{a\Omega}$$

where N is the BRUNT-VAÏSALA frequency, D the thickness of the troposphere, a the radius and Ω the angular velocity of the planet. For Earth, the critical shear is about $U \simeq 30$ m sec^{-1} for D = 10 km, a zonal wind velocity very easily reached in the laminar HADLEY circulation (see section 3). This is why the atmospheric flow is constantly producing large quasi-horizontal waves which interact through the non-linear terms of the governing equations and produce in turn all scales of motion (Fig.4 and 6). Thus, the planetary atmosphere circulation as well as oceanic currents, are fully turbulent flows involving all scales of eddies from the large cyclonic perturbations (1000-2000 km) to mesoscale motions (10-100 km) and even microscale turbulence (1 mm-100 m).

6. ATMOSPHERIC CIRCULATION AS TWO-DIMENSIONAL TURBULENCE

For all but the smaller scale eddies (100 km or less), vertical stability imposes a very marked anisotropy, so much anisotropy in fact, that these perturbations may be considered as two-dimensional, non-divergent eddies (CHARNEY, 1971). This anisotropy has a very definite influence on the spectral distribution and the transfer of kinetic energy between different scales (LEITH, 1971). Keeping to essentials, we may rewrite the dynamic equations (4.14) for two-dimensional viscous flow:

$$\frac{\partial u_i}{\partial t} + u_j \frac{\partial u_i}{\partial x_j} + \frac{\partial}{\partial x_i}\left(\frac{p}{\rho_s}\right) = \nu \frac{\partial^2 u_i}{\partial x_j \partial x_j} \qquad (6.1)$$

where $i,j = 1,2$ and where summation over repeated indices is implied. Taking the curl of (6.1) and using the continuity equation:

$$\partial u_i / \partial x_i = 0$$

the following equation for the vorticity $\zeta = \text{curl } \vec{V}$, obtains:

$$\frac{\partial \zeta}{\partial t} + u_j \frac{\partial \zeta}{\partial x_j} = \nu \frac{\partial^2 \zeta}{\partial x_j \partial x_j} \qquad (6.2)$$

Multiplying (6.1) and (6.2) by u_i and ζ respectively and integrating over the whole domain occupied by the fluid, i.e. a spherical shell around the planet, yields the dissipation rates of total kinetic energy K and total square vorticity or enstrophy Z:

$$\frac{\partial K}{\partial t} = \frac{\partial}{\partial t}\left[\frac{u_i u_i}{2}\right] = -\nu\left[\frac{\partial u_i}{\partial x_j} \cdot \frac{\partial u_i}{\partial x_j}\right] = -\epsilon \qquad (6.3)$$

$$\frac{\partial Z}{\partial t} = \frac{\partial}{\partial t}\left[\frac{\zeta^2}{2}\right] = -\nu\left[\frac{\partial \zeta}{\partial x_j} \cdot \frac{\partial \zeta}{\partial x_j}\right] = -\eta \qquad (6.4)$$

where the brackets $\begin{bmatrix} \\ \end{bmatrix}$ indicate the result of integration over the whole sphere. These relations show that, except for viscous dissipation, both K and Z are <u>integral invariants</u> in two-dimensional non-divergent flows. We see further that ε is proportional to νk^2 for wave vector k, while η is proportional to νk^4 and must dominate in the short wavelength limit of the spectrum. Thus, if the short wavelength tail of the energy spectrum is a steady inertial subrange, it must be determined by the <u>enstrophy dissipation rate</u> η only. Note that the dimension of η is T^{-3}. We can deduce then, on dimensional grounds, that the short wavelength cut-off of the energy density spectrum must be sharper than the usual -5/3 law, to wit:

$$E(k) \simeq A\eta^{2/3} k^{-3} \tag{6.5}$$

This k^{-3} dependence has indeed been found by numerical simulation of non-divergent two-dimensional flows with high enough REYNOLDS number (FOX and LILLY,1971), while a fairly consistent body of evidence points out to the same property for the Earth atmosphere general circulation as shown in Figure 20 (WIIN-NIELSEN(1967); JULIAN et al. (1970); KAO and WENDELL (1970)). It can be seen furthermore that the two integral invariants K and Z determine the mean wavenumber of the large scale quasi-horizontal waves. Assume for example that the (non-divergent) flow is represented by a stream function Ψ which can be expanded in series of spherical harmonics Λ_n, i.e. eigenfunctions of the Laplace equation on the sphere:

$$\nabla^2 \Lambda_n = - \lambda_n \Lambda_n$$

with eigenvalues $\lambda_n = n(n+1)/a^2$ roughly equal to the square of the wavenumber n. Then, it is easily found that:

$$2K = \left[(\nabla \Psi)^2 \right] = \sum_n \lambda_n \left[\Lambda_n^2 \right]$$

$$2Z = \left[(\nabla^2 \Psi^2)^2 \right] = \sum_n \lambda_n^2 \left[\Lambda_n^2 \right]$$

so that a mean wavenumber $\bar{n}$ or mean eigenvalue $\bar{\lambda}$ is given by:

$$\bar{\lambda} = \frac{\sum_n \lambda_n^2 \left[\Lambda_n^2\right]}{\sum_n \lambda_n \left[\Lambda_n^2\right]} = \frac{Z}{K}$$

Thus, it is not enough that a numerical computation scheme conserves the total kinetic energy K. If spurious, systematic creation of enstrophy Z is allowed, the mean eigenvalue $\bar{\lambda}$ may creep up, and more and more energy could flow in the higher wavenumbers leading to novel kind of "non-linear" computational instability (PHILLIPS, 1959). This instability is indeed checked by enstrophy conserving computation schemes (ARAKAWA, 1966); (SADOURNY etal. 1968).

A further (basic) difficulty results from the fact that the spectrum of atmospheric motions extends to exceedingly small wavelengths. No numerical computation can account for this wide variety of scale: simulated geophysical flows are by necessity truncated at some finite wavenumber k_* as shown in Fig. 7: small scale (sub-grid scale) motions cannot be explicitly taken into account although they do exist in real geophysical flows and do interact significantly with larger scale features. This is the one fundamental irreducible difference between "theoretical" or simulated flows and the real thing. A variety of closure hypotheses have been proposed for estimating the average energy dissipation resulting from the interaction of explicitely computed spectral components with the sub-grid scale motions (LILLY, 1951). The most practical approximation is based on the recent concept of two-dimensional turbulence and consists in replacing the transfer term T(k) of the spectral energy density equation:

$$\frac{d\,E(k)}{dt} = T(k) - 2\,\nu k^2\,E(k) \qquad (6.7)$$

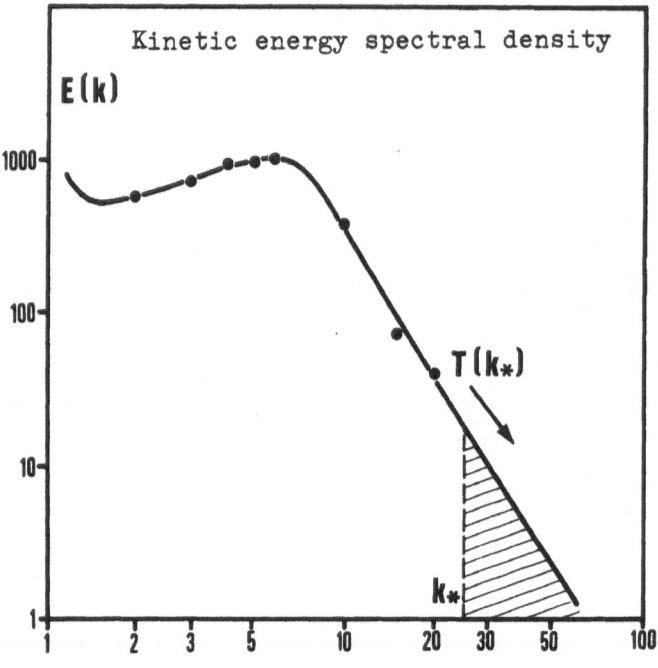

Figure 7
Wave vector **k**

by an appropriate eddy viscosity dissipation:

$$T(k_*) \simeq - 2\nu_* k_*^2 \quad E(k_*) \qquad (6.8)$$

where the eddy viscosity coefficient ν_* is not constant but
increases where the vorticity gradient is strong and also depends
upon the cut-off wavevector k_* (LEITH, 1971):

$$\nu_* \simeq B \ |\nabla\zeta| \ k_*^{-3} \qquad (6.9)$$

This closure hypothesis (6.8 - 6.9) accounts quite well for the
mean energy and enstrophy dissipation in quasi two-dimensional
flows and do preserve the proper spectral shape (6.5). But it does
not account for the detailed dynamic interaction between the
simulated flow and sub-grid scale eddies : This interaction exists

however in the real fluid and constitutes as much _error_ source
in the simulation of the flow dynamics. This error is initially
restricted to the shorter wavelengths but it does progressively
invade the whole spectrum so that the computed flow becomes
eventually as different from the real flow as two randomly chosen
states of motion would be. In this sense, a geophysical flow is
not _indefinitely_ predictable. For the Earth atmosphere and in
the current state of computational resolution and observing
techniques, this basic predictability limit may be of the order
of 10 days or so (LORENZ, 1969);(LEITH, 1971). This fundamental
spectrum truncation error is not the limiting factor in current
atmosphere models, however, as many numerical experiments have
shown: forecasting accuracy is not materially improved when the
spatial resolution of the computation is further reduced beyond
100 km or so (BAER and ALYEA, 1971; MIYAKODA et al., 1971; WELLCK
et al., 1971).

Closure hypotheses used to estimate the dynamic effects of
sub-grid scale eddies in term of explicitly computed spectral
components of the motion are a first example of the _parameteriza-
tion of sub-grid scale processes_ which account for the main
atmospheric energy sources and sinks. This whole subject of small
scale processes parameterization may be the key to any real
understanding of climatic changes, more so than the more obvious
computational problems which seem to be well understood now as
we shall see presently.

PART TWO

7. FORMULATION OF THE NUMERICAL PROBLEM

The simulation of the dynamics of atmospheric circulation is
based on the numerical integration of a set of governing equations,
usually the "primitive equations" (4.15), under hopefully realistic
conditions i.e. taking into account boundary conditions like the
topography of the Earth surface and primary as well as secondary
energy sources or sinks like solar radiation, long-wave radiation,
evaporation, latent heat of condensation ... Now, the governing
equations include only first order time derivatives of the (four-
dimensional) field $\Psi(u,v,\theta,\rho)$ which describe the state of motion
of the fluid. They may thus be written symbolically:

$$\frac{\partial \Psi}{\partial t} = H \ \Psi(t) \qquad\qquad (7.1)$$

where H is a differential operator involving space derivatives of
the Ψ-field at one time t only. In view of the aperiodic nature
of the solutions (a turbulent flow never exactly repeats itself),
it is best to consider the integration of (7.1) as an <u>initial
value problem</u>, i.e. compute the values of Ψ stepwise at times Δt,
$2 \Delta t, \ldots n\Delta t$ starting from a known or assumed value at time
t = 0. This process is equivalent of course to approximating
equation (7.1) by a finite difference analogue like:

$$\frac{\Psi(t+\Delta t) \ - \ \Psi(t)}{\Delta t} = H \ \Psi(t) \qquad\qquad (7.2)$$

for example. Note that this particular first order forward
difference approximation is not applicable to our problem as it
is uncondionally unstable and causes the approximate solution to
oscillate wildly about the real value. The second order "leapfrog"

approximation is quite generally used instead:

$$\frac{\Psi(t+\Delta t) - \Psi(t-\Delta t)}{2\Delta t} = H \ \Psi(t) \qquad (7.2)$$

which allows $\Psi(t+\Delta t)$ to be estimated from the known values of the
field on two successive time steps $t - \Delta t$ and t. The leapfrog
time integration scheme has, in addition to second order accuracy
and simplicity, the advantage of neither damping nor amplifying
spuriously the wave like disturbances of the basic flow. It does,
on the other hand require very short time steps ($c_o \Delta t < \Delta x$) for
stability so that more complicated implicit numerical schemes are
preferred in some instances. The advantages, and drawbacks of
these various possible schemes are well known and have been
excellently discussed; see RICHTMYER and MORTON (1967), FISCHER
(1965), LILLY (1965), KURIHARA (1965).

Now we come to the delicate and not at all trivial
problem of choosing a spatial representation of the field
$\Psi(x,y,z,t)$ suitable for numerical computation of $H\Psi$ in equation
(7.1). Since a turbulent flow contains all scales of motions, i.e.
an infinite number of degrees of freedom on the one hand, and a
computer memory contains only a finite number of words on the
other hand, the process of numerical integration involves necessarily a truncation of the allowed spectrum of motion. There are
basically two ways in which this truncation can be done: replacing
Ψ by a suitably truncated (finite) expansion in series of some
orthogonal functions like spherical harmonics, or sampling the
continuous field $\Psi(x,y,z)$ to make a three-dimensional array of
discrete values $\Psi(x_i \ y_i \ z_i)$. The first approach is known as the
spectral integration method while the second is the finite
difference method.

A straightforward application of the spectral method would
not do for high spatial resolution models of the atmospheric
circulation, because the non-linear horizontal advection operator
in the governing equation (4.15) causes $H\Psi$ to contain two-
harmonics interaction terms. The method requires therefore

summing a series of four indices interaction products like:

$$\sum_l \sum_m \sum_{l'} \sum_{m'} \quad Y_{lm}(\theta,\phi) \quad Y_{l'm'}(\theta',\phi') \quad C(l,m,l',m')$$

and expressing these products in term of $Y_{l+l',p}(\theta,\phi)$. Now the number of such products increases like the fourth power of the spatial resolution or the fourth power of the limiting wavevector k_* , a distinct disadvantage considering the number of arithmetic operations required (the number of operations increases like the number of points in the grid, i.e. like the second power of the resolution with the finite difference method).

The spectral method is nevertheless quite applicable and precise, when a very high spatial resolution is not required like short term global weather forecast (ELLSAESSER, 1966). Furthermore even when excellent resolution is required, a very attractive mixed method has been proposed by ORSZAG (1969,1971) combining some advantages of the spectral and the discrete grid-representa- tions. This approach is basically a spectral expansion method where however, the interaction (non-linear) terms are computed in finite difference form in ordinary space as the Fast Fourier Transform or similar algorithms allows rapid transformations from a truncated spectral series in k-space to discrete grid-values in ordinary space. This new method will be discussed elsewhere. We shall then devote the rest of the paper to the more commonly used finite difference methods.

8. LOCAL CURVILINEAR COORDINATES VERSUS PLANE MAPPING

The alternative to spectral methods consists in replacing
the continuous fields which appear in equation (7.1) by a set of
discrete field values specified on a three dimensional array or
grid covering the domain under consideration. In view of the
technical difficulties attendant to using local coordinates on a
curved (spherical) surface, it seems appropriate to map this
surface onto a plane and utilize cartesian geometry thereafter.
This is straightforward enough for simulating the atmospheric
circulation over one hemisphere when a simple boundary condition
at the Equator can be assumed,like mirror-symmetry or frictionless
slip flow. The best procedure would then be to apply the (conform-
al) stereographic projection for mapping the hemisphere onto a
circular area and sample the field in this area at the grid points
of a square or otherwise regular array (Figure 8). It is equally

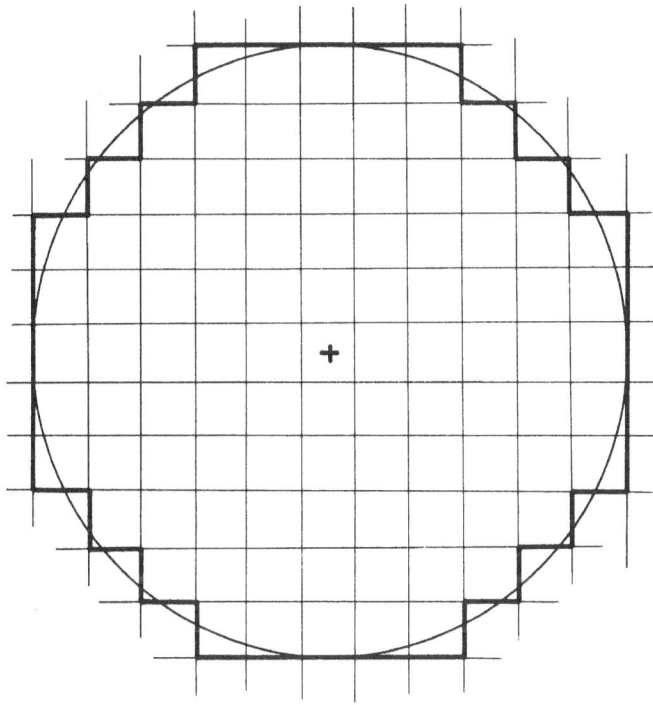

Figure 8
A square array (x,y) on the
stereographic projection of one hemisphere

possible to map a spherical surface onto a plane with the
exception of two polar caps, using the Mercator projection or any
similar longitude-latitude mapping. All such mapping techniques
suffer however from one basic deficiency : a singularity at the
poles which makes it quite difficult if not outright impossible
to design consistent differencing schemes in the whole domain.
Thus, it would seem attractive to develop a local approach, i.e.
approximating the spherical surface by a polyhedron with a great
many faces. Each face is then a small, irregular polygonal
surface tangent to the sphere where all vector and scalar
quantities can be properly defined in term of a local coordinate
system. Several attempts have been made in this direction with
a fair amount of success (KURIHARA, 1965; KURIHARA and HOLLOWAY,
1967; SADOURNY, ARAKAWA and MINTZ, 1968; WILLIAMSON, 1970). But
approximating differential operators with finite differences on
an irregular array necessarily results in first order truncation
errors even though the differencing schemes may be everywhere
consistent and may conserve the total mass, energy and enstrophy
of the flow. SADOURNY has demonstrated that under these circumstances, the first order truncation errors of the local approach
exceed the detrimental effect of the singular regions found in
the mapping approach (SADOURNY, 1972). We believe therefore that
any formulation of the primitive equations in finite differences
should start by mapping the spherical domain of the flow onto one
plane map (Mercator or similar projection) or more conveniently
several connected planar maps. This last approach is the one
chosen by Laboratoire de Météorologie Dynamique : it consists in
a central projection of the spherical surface onto the 20 connected triangular faces of the icosaedron (Figure 9).

NORTH POLE

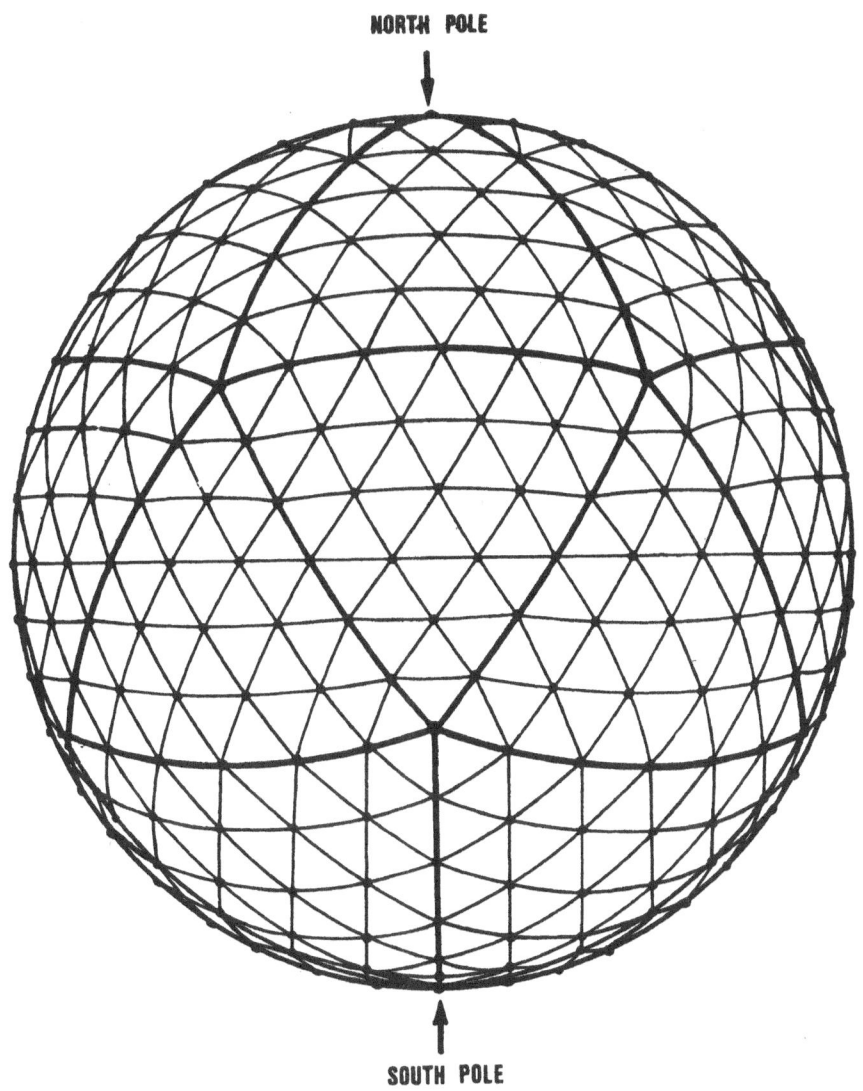

SOUTH POLE

Figure 9
Mapping the sphere onto the icosaedron

Let us then refer to equations (5.1) for the barotropic flow of an incompressible atmosphere and write these equations in a slightly different form, more convenient for the purpose at hand:

$$\frac{\partial \vec{V}}{\partial t} + (f+\zeta)\ \vec{k}\ x\ \vec{V} + grad(\phi + \frac{1}{2}V^2) = 0 \qquad (8.1)$$

$$\frac{\partial \phi}{\partial t} + div\ (\phi\vec{V}) = 0 \qquad (8.2)$$

where again $\vec{V}$ is the horizontal velocity and $(f+\zeta)$ is the absolute vorticity of the flow on the rotating planet. Now, grad, div and $\zeta = curl\ \vec{V}$ are two dimensional differential operators to be expressed in a covariant form, using the metric coefficients g_{ij} appropriate to the particular mapping of the spherical surface. Introducing:

$$G = \left[\det\ g_{ij}\right]^{1/2}$$

and the antisymmetrical tensor :

$$\varepsilon_{ij} = \varepsilon^{ij} = \begin{vmatrix} 0 & 1 \\ -1 & 0 \end{vmatrix} ,$$

we find that equations (8.1)(8.2), become:

$$\frac{\partial u_i}{\partial t} + \varepsilon_{ij}\ u^j(Gf + \varepsilon^{km}\frac{\partial u_m}{\partial x^t}) + \frac{\partial}{\partial x^i}(\phi + \frac{1}{2}u_j\ u^j) = 0 \qquad (8.3)$$

$$G\frac{\partial \phi}{\partial t} + \frac{\partial}{\partial x^i}(G\phi u^i) = 0 \qquad (8.4)$$

where x^1 ,x^2 are the coordinates of a local <u>cartesian</u> system and u^1,u^2, ϕ are three coupled fields to be determined simultaneously.

9. INTRODUCTION TO THE PROBLEM OF SPACE DIFFERENCING

The first thing then is to select one particular (regular) array R of points which would best serve the conflicting purposes of simplicity, accuracy and good spectral resolution. The simplest choice is evidently the square grid depicted in Figure 10, where all three dynamic fields are sampled at the same points.

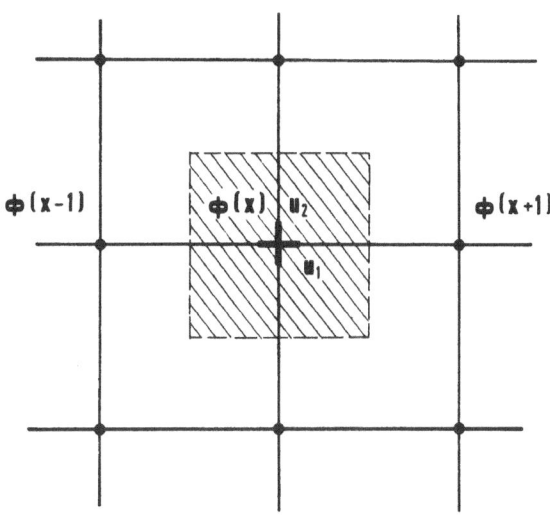

Figure 10
The simple square grid

Obviously, this scheme does not provide the best spectral resolution because the acceleration $\partial u/\partial t$ at grid point x for example will be associated with the centered approximation of the gradient of $\phi (\Delta x = 1)$:

$$\frac{\partial u(x)}{\partial t} \sim \frac{\phi (x+1) - \phi (x-1)}{2} \qquad (9.1)$$

Thus, waves longer than 4 Δx which produce a significant gradient over a distance 2 Δx, are properly integrated under this scheme. Grossly underestimated phase velocities result for shorter wavelengths. In particular, stationary "checker-board" perturbations with a wavelength 2 Δx may cause large geopotential variations from one grid point to the next and yet induce no motion of the fluid since:

$$\phi(x-1) = \phi(x+1) = \phi(x+3) = \ldots$$

$$\phi(x) = \phi(x+2) = \phi(x+4) = \ldots$$

A better arrangement then, is to use a <u>staggered grid</u> consisting in two independent arrays R' and R", where scalar quantities such as the geopotential ϕ , vorticity ζ and thermo-dynamic parameters are defined on the first array and vector components u^1, u^2 on the second array (Figure 11). It can be seen

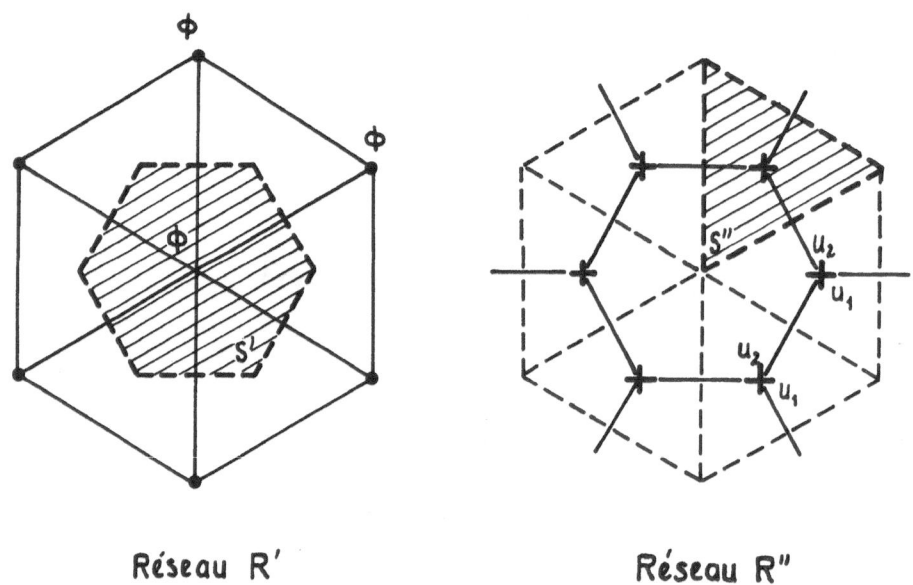

Réseau R'   Réseau R"

Figure 11
Example of a staggered-grid
with two sub-arrays R' and R"

readily that stationary patterns can still exist but with a shorter
wavelength (i.e. $\sqrt{2}$ Δx for the square grid). A further improvement
in this direction can be achieved by introducing three different
arrays R',R" and R"' for scalar quantities, and each component
u^1 and u^2 of vector quantities, respectively (Figure 12). There,
no stationary pattern remains since any gradient of ϕ between
adjacent grid points produce a compensating flow.

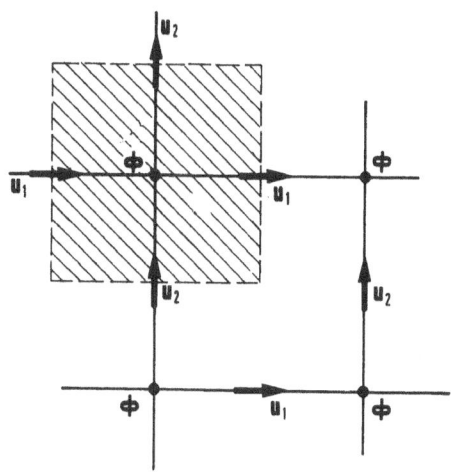

Figure 12
Staggered grid with three sub-arrays

This last arrangement is not applicable to geophysical flows
on a rotating planet, however, because equations (8.1) include a
Coriolis acceleration term which amounts to a rotation of the
velocity vector in the horizontal plane. It is thus necessary
that <u>the two components</u> u^1,u^2 of the velocity be specified at
the same place if the Coriolis acceleration is to be computed
accurately (the truncation errors involved in using interpolated
values would be excessive). Further, specifying complicated

lateral boundary conditions may prove to be intractable with a
staggered grid. This is why the simpler non-staggered grid
(Figure 10) is generally preferred for the simulation of marine
currents with irregular bottom topography (BRYAN, 1969). For the
same reason, a non-staggered triangular grid is used for global
circulation simulation at Laboratoire de Météorologie Dynamique
to make possible a consistent and accurate connexion of adjacent
maps of the spherical domain.

Let us then introduce, for generality, two regular arrays
R' for scalar quantities (such as ϕ, curl $\vec{V}$, G, etc...) and R"
for vector components (u^1, u^2), with the understanding that R'
and R" may coincide if it is so decided. Further, let us
introduce two scalar products for scalar quantities (ϕ, Ψ) and
vector components (u, v) respectively:

$$< \phi, \Psi > = \sum_{R'} \phi \Psi s' \qquad (9.2)$$

$$\ll u, v \gg = \sum_{R''} u v s'' \qquad (9.3)$$

where s' and s" are the areas of the cells associated with each
R' and R" grid point, respectively (Figure 11). Note that the
ensemble of all cells s' as well as the ensemble of all s" add
up to the total area of the sphere. Finally, we note than the
governing equations (8.3) (8.4) involve three kinds of differential
operators associated with grad, div and curl, respectively, as well
as products of scalar and vector quantities. We need therefore:

- One <u>interpolation operator</u> A', i.e. a linear application
  of R" onto R' for computing the (interpolated) value of
  a vector quantity on R'. For example:

$$A'(u) = \sum_{\xi} a'(\xi) u(x' + \xi) \qquad (9.4)$$

where the sum is extended to any number of grid points
x'+ ξ of the R" array in the vicinity of x' (depending
upon the order of the approximation).

- The corresponding interpolation operator A" for computing
the value of a scalar field on R" :

$$A" (\phi) = \sum_{\xi} a"(\xi) \phi(x"+\xi) \qquad (9.5)$$

- One <u>differencing operator</u> δ'_j for each direction (j=1,2)
applicable to vector quantities for computing the (scalar)
divergence field on R'.

- One differencing operator $\delta"_j$ (j=1,2) applicable to
scalar quantities for computing the (vector) gradient
field on R".

- Another differencing operator Δ'_j (j=1,2) applicable again
to vector quantities for computing the (scalar) vorticity
field on R'.

All three differencing operators are linear applications
like (9.4) or (9.5) as the case may be, with a different set of
coefficients, of course. Now, the finite difference approximations
of equations (8.3) and (8.4) read:

$$\frac{\partial u_i}{\partial t} + \epsilon_{ij} u^j A" \left[Gf + \epsilon^{km} \Delta'_k u_m \right] + \delta"_i \left[\phi + \frac{1}{2} A'(u_k u^k) \right] = 0 \qquad (9.6)$$

$$G \frac{\partial \phi}{\partial t} + \delta'_i \left[A"(G\phi) u^i \right] = 0 \qquad (9.7)$$

while the total kinetic energy is:

$$E = \frac{1}{2} < G\phi , \phi + A'(u_k u^k) > \qquad (9.8)$$

Differentiating with respect to time yields:

$$\frac{dE}{dt} = \ <\ G\ \frac{\partial \phi}{\partial t}, \phi\ +\ \frac{1}{2}\ A'(u_k u^k)>\ +\ <\ G\phi,\ A'(u^k\ \frac{\partial u_k}{\partial t})>$$

$$=\ <\ G\ \frac{\partial \phi}{\partial t}, \phi\ +\ \frac{1}{2}\ A'(u_k u^k)>\ +\ \ll A''(G\phi) u^k,\ \frac{\partial u_k}{\partial t}\ \gg$$

on the condition that the interpolation operators A' and A" be
symmetrical (see 9.11). Making use of the governing equations
(9.6) - (9.7) and seeing that the antisymmetrical term (ε_{ij}) does
not contribute to the sum, we find:

$$-\frac{dE}{dt} = \ <\delta'_i\ \left[A''(G\phi)\ u^i \right],\ \ \phi\ +\ \frac{1}{2}\ A'(u_k u^k)>$$

$$+\ \ll A''(G\phi)\ u^i,\ \ \delta''_i\ \left[\phi\ +\ \frac{1}{2}\ A'(u_k u^k) \right]\ \gg \qquad (9.10)$$

The right hand side of (9.10) will vanish, and thus, exact
conservation of the total kinetic energy of this barotropic flow
will ensue if only the interpolation operators A', A" are symmetrical and the differencing operators δ'_i, δ''_i antisymmetrical:

$$<\ \phi\ ,\ A'u\ >\ =\ \ll A''\phi, u \gg \qquad (9.11)$$

$$<\ \delta'_i\ u, \phi>\ =\ -\ \ll u, \delta''_i\ \phi \gg \qquad (9.12)$$

Note that this last condition is nothing but the finite difference
analogue of the familiar differential relation, valid on a
spherical domain S:

$$\iint_S (\phi \ \text{div} \ \vec{V} + \vec{V}. \ \overrightarrow{\text{grad}}\phi) \ dx_1 \ dx_2 = \iint_S \text{div}(\phi\vec{V})dx_1 \ dx_2 = 0$$

It is straightforward to prove that the exact conservation of total mass also ensues from condition (9.12). From (9.7):

$$< G \ \frac{\partial \phi}{\partial t} \ , \ 1 > = - < \delta'_i \ A''(G\phi)u^i \ , 1 >$$

$$= \ll A''(G\phi)u^i, \ \delta''_i(1) \gg \equiv 0$$

The third differencing operator Δ'_i related to curl $\vec{V}$ may or may not be chosen identical to δ'_i, depending upon further dynamical constraints (conservation of total enstrophy Z) and the nature of the lateral boundaries. For a more detailed discussion of this point, see SADOURNY (1972).

A very simple example of an energy and mass conserving difference scheme is the "box-method" based on a single square grid (R' ≡ R") like Figure 10, and identical differencing operators:

$$\delta'(u) \ = \ \frac{u(x+1) \ - \ u(x-1)}{2}$$

$$\delta''(\phi) \ = \ \frac{\phi(x+1) \ - \ \phi(x-1)}{2}$$

Then:

$$< \delta'(u), \phi> + \ \ll u, \delta''(\phi) \gg$$

$$= \ \sum_x \frac{u(x+1) \ - \ u(x-1)}{2} \ \phi(x) \ + \ u(x) \ \frac{\phi(x+1) \ - \ \phi(x-1)}{2}$$

vanishes evidently since any product $u(x)\phi(x+1)$ appears twice in the sum, with opposite numerical coefficients $+1/2$ and $-1/2$.

10. RESULTS

It may be taken from the above presentation that the problems
involved in the numerical integration of the "primitive equations"
of geophysical fluid dynamics, are well understood if not completely in hand. A rational basis for choosing one representation of
the dynamic fields $\phi, u^1, u^2(x,y,z)$ has been laid and the symmetry
(or anti-symmetry) property of the finite difference operators
required to exclude the spurious generation (or loss) of mass,
kinetic energy, enstrophy... have been formulated. We know why
these conservation properties are essential to prevent gross
misrepresentations of the flow patterns and spectral characteristics such as the spurious transfer of kinetic energy from long
to short wavelengths, known as the "non-linear instability"
(PHILLIPS, 1959). Since all these primary difficulties have
been overcome, we may guess that all general circulation models
now in operation are at least fairly successful in simulating,
and even forecasting the medium-term dynamic evolution of the
atmosphere!

This is not to say that any finite difference integration
scheme is very accurate in the shorter wavelength limit of the
truncated spectrum of motions. In fact, quite the opposite is
true: no second order space differencing scheme ever yield
accurate dynamical results for waves shorter than 4 grid intervals Δx, say. An illustration of this fact may be found in
comparing to the analytic solution (5.3), the frequency dispersion
relation $\omega = \omega(k)$ obtained with various finite difference
approximations of the linearized equations of motion (5.2) :
Figure (13) through (15). Note that for both the geostrophic
(j=0) and gravity (j = $\pm$ 1) modes, the phase velocity found
for the finite difference approximate solution becomes much too

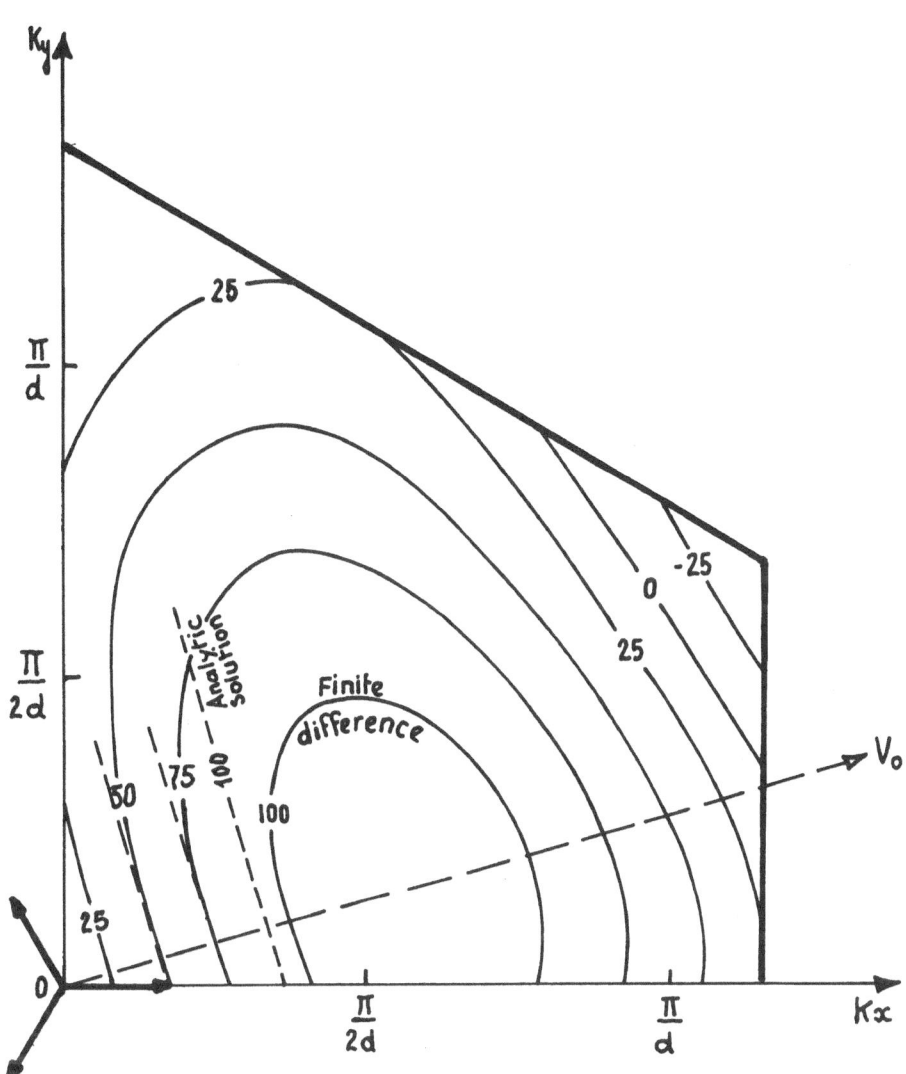

Figure 13
Isolines of the frequency $\omega = \omega\,(k_x, k_y)$ for
finite difference solutions corresponding
to the geostrophic (j=0) mode (hexagonal grid)

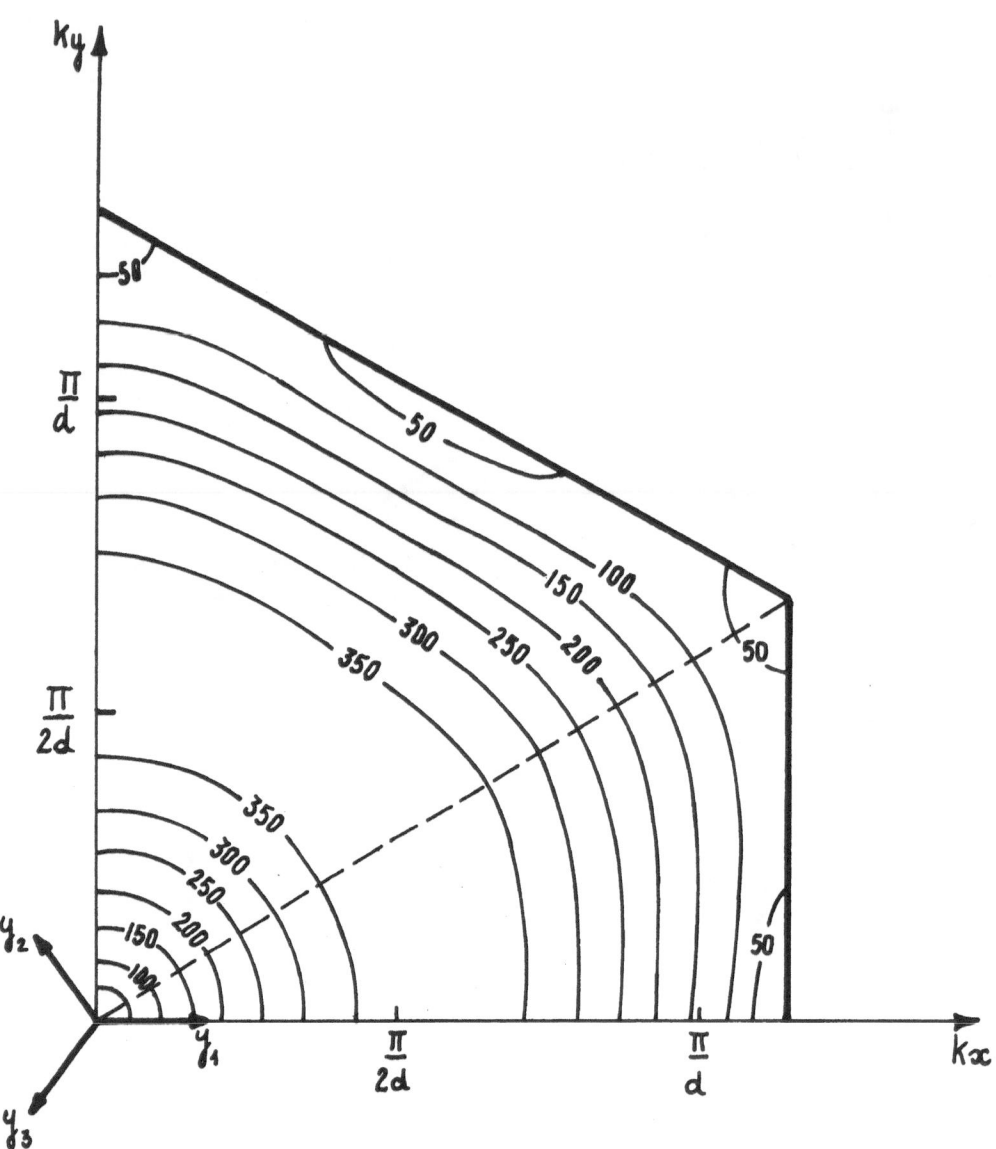

Figure 14
Same as Fig. 13 for gravity waves
$(j = \pm 1)$

53

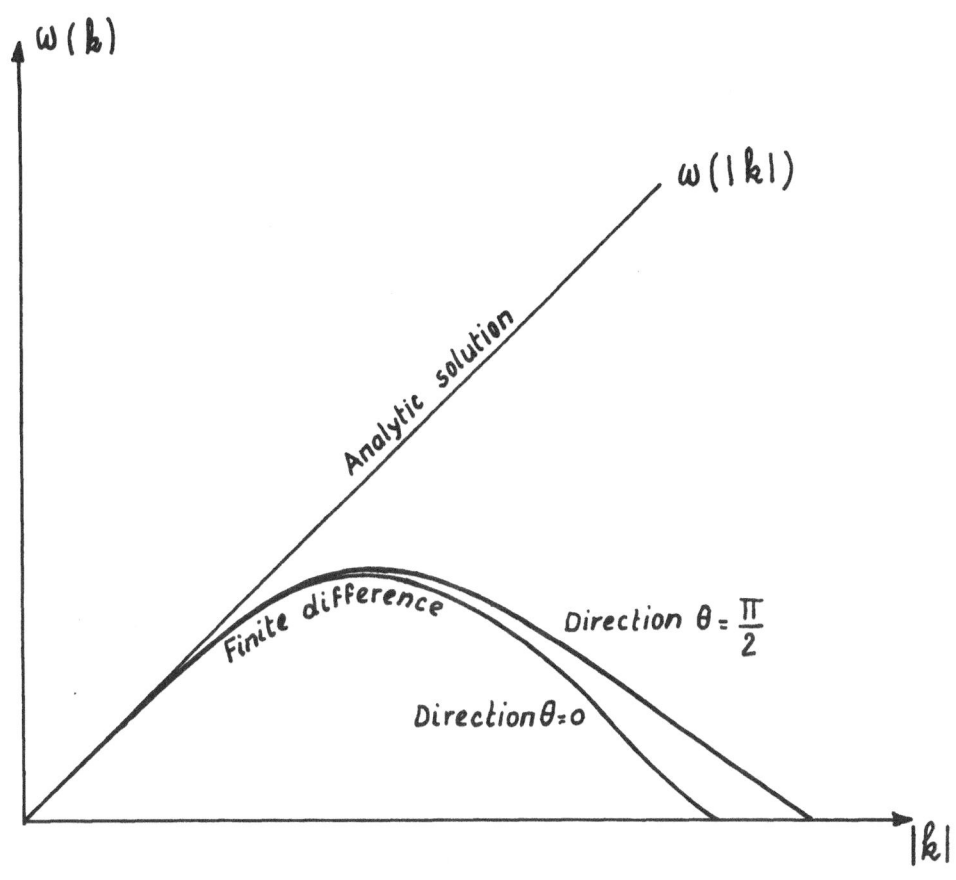

Figure 15
Dispersion relation ω(k) for finite
difference solutions corresponding to the j = 0 mode
(hexagonal grid)

small for wavevectors larger than $k_c/2$ roughly. An improved
accuracy obtains with higher order space differencing schemes
involving a larger number of grid points but it is not clear
that such more complicated differencing schemes are economical
when compared to straightforward reduction of the mesh size Δx.
Present day computers make it possible to carry the numerical
integration of the primitive equations (4.15) with an effective
mesh of 250 km corresponding to 6000-10.000 grid points over
the globe or even 125 km, corresponding to four times as many
grid points. Even so, numerical truncation errors are still
believed to be the main cause of discrepancy in extended
(3-5 days) numerical weather prediction. Numerical experiments
designed to test the effect of grid resolution have indeed
demonstrated a significant error reduction when using a finer
grid (MIYAKODA et al., 1971; WELLCK et al., 1971). It seems
likely on the other hand, that phase velocity errors for the
shorter wavelength motions are not very detrimental to simulations of the mean dynamical behavior of geophysical flows, as
evidenced by the remarkable verisimilitude of simulated climate
for the planet Earth, not only regarding the average temperature
or velocity fields but also the second moments like eddy momentum
and energy transfer.

Figure 16, taken from MANABE, SMAGORINSKY and STRICKLER
(1965), shows computed mean vertical temperature profiles for
various latitudes, compared to observed values. Actually, the
accuracy of these profiles speaks more in favor of these authors'
computation of the vertical radiative transfer than anything else
since the mean temperature distribution computed for a static
atmosphere (order zero approximation) is changed only slightly
by horizontal exchanges.

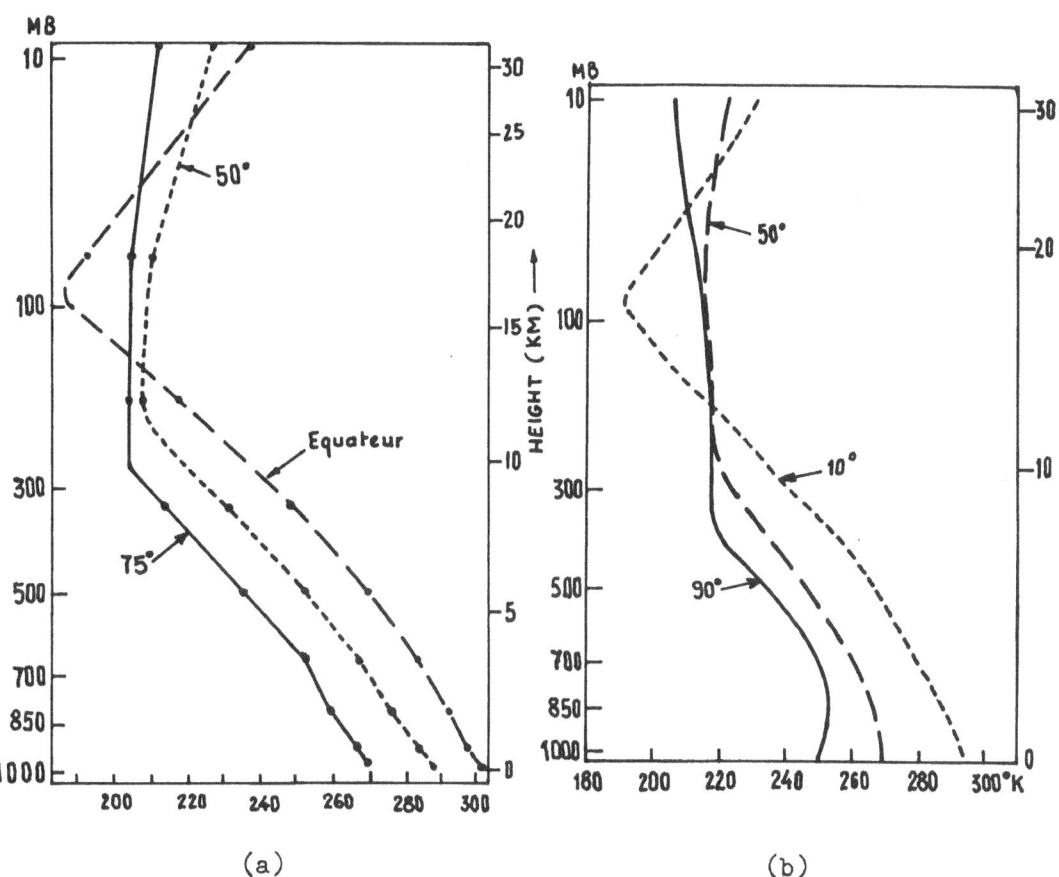

Figure 16
Computed (a) and observed (b) mean temperature
profiles (degrees Kelvin) at various latitudes

56

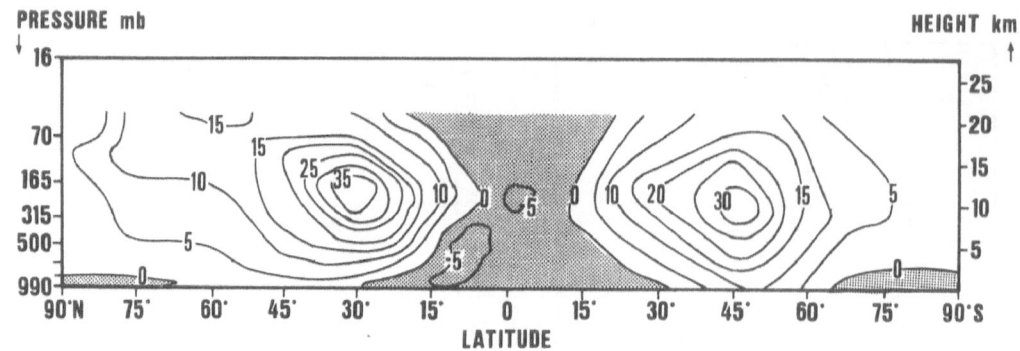

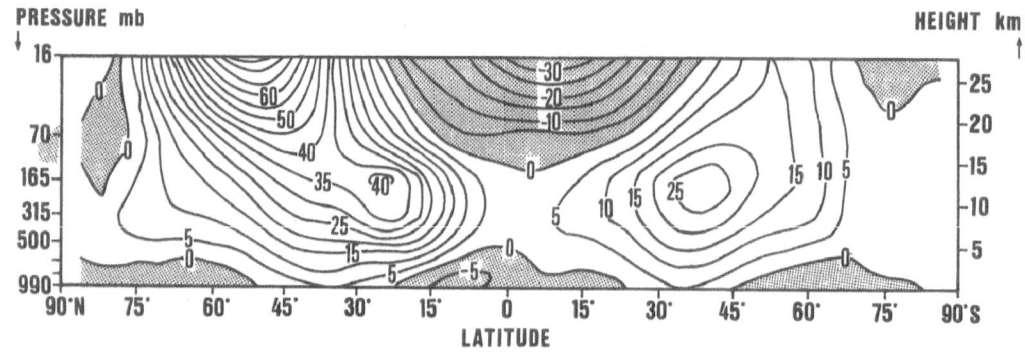

Figure 17
Latitude height distribution of the mean zonal
wind (m sec -1) from observations "a" and one
numerical simulation "b"

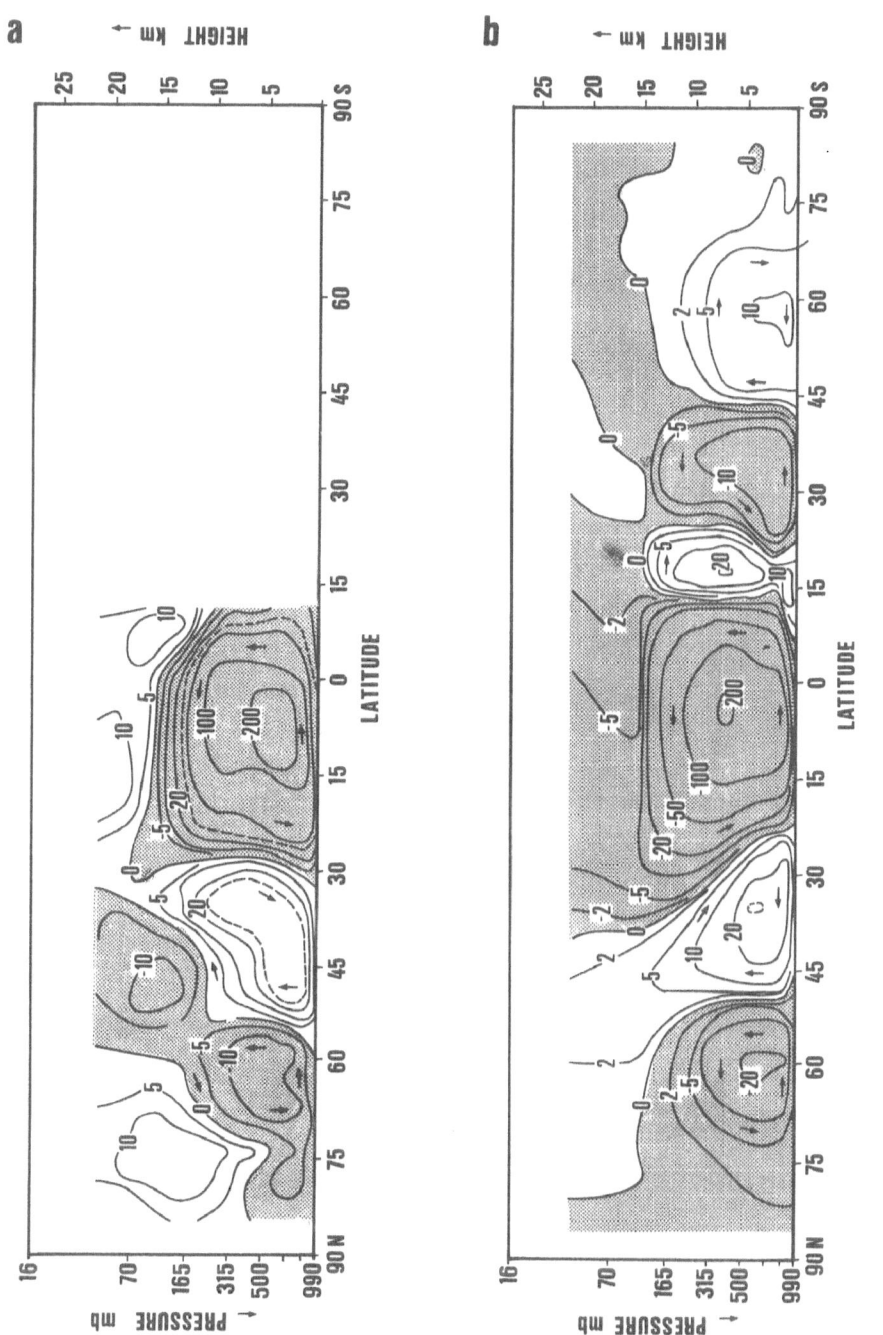

Figure 18
Stream function of the mean meridional circulation
from observations "a" and numerical simulation "b"

A more critical test, taken from MANABE, HOLLOWAY and STONE (1970), is a comparison of the mean zonal (Figure 17) and vertical-meridional flows (Figure 18). Note that the strengths of the (direct) tropical HADLEY circulation and the indirect FERREL circulation in the meridional plane are quite well reproduced in the troposphere, if not quite in the stratosphere.

Finally, the more sophisticated general circulation models can also simulate quite well the transient behavior of the atmosphere and generate the proper amount of eddy transport of angular momentum towards the pole, for example (Figure 19 taken from MANABE, SMAGORINSKY and STRICKLER, 1965). They do also provide a fairly consistent, if not completely satisfying account of the interaction between different spectral components of the motion, as evidenced by the very accurate spectral distribution found for the eddy kinetic energy (Figure 20, taken from MANABE, SMAGORINSKY, HOLLOWAY and STONE, 1970).

We may conclude then that numerical models of the atmospheric circulation have now reached a degree of sophistication and verisimilitude such that they may be used to investigate quite intricate dynamic phenomena which could not be approached by observation only.

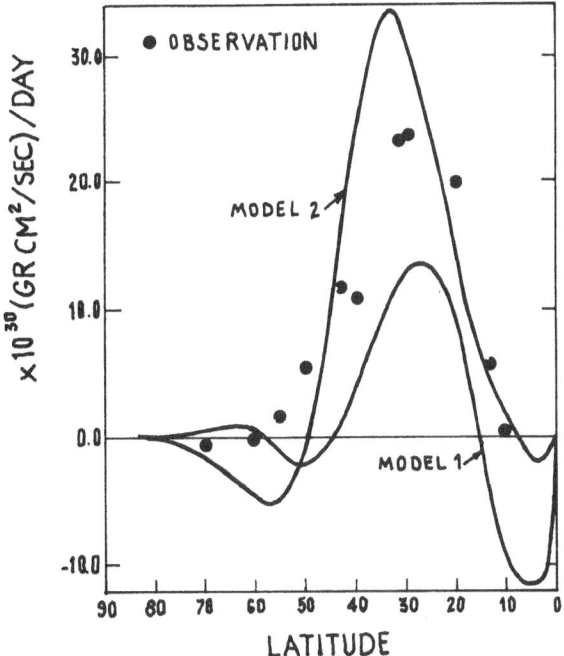

Figure 19

Poleward eddy-transfer of angular momentum
in various numerical simulations of the
general circulation

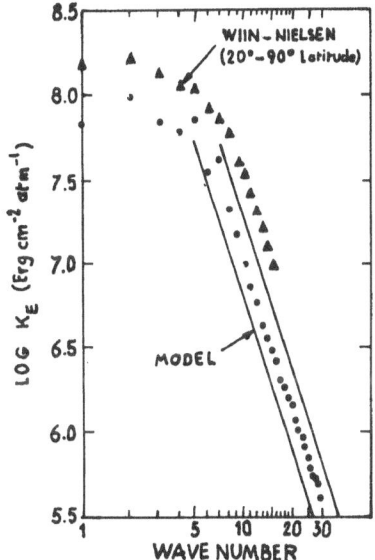

Figure 20

Eddy kinetic energy spectrum of the real atmosphere (▲)
and a numerical simulation of the general circulation (•)

PART THREE

11. FUTURE PROSPECTS

Taking the broad view, one could say that most if not all
major difficulties involved in extended range (one week or more)
numerical weather forecasting are solved in principle. Numerical
techniques have indeed evolved to the point that consistent
and reasonably accurate computing schemes are available. A much
deeper understanding of the atmosphere quasi-geostrophic perturbations have led to the concept of two-dimensional turbulence
and appropriate closure hypotheses allowing the turbulent motion
of the atmosphere to be reduced to truncated representations with
large but finite numbers of degrees of freedom. The Global
Atmospheric Research Programme is now set to iron out the remaining discrepancies and obtain empirical laws for the significant
energy sources and sinks in the Earth atmosphere.

But deterministic numerical weather forecasting is not the
ultimate goal and may even be considered a simple problem in
principle, if not in actual practice, because it must necessarily
be restricted to a fairly short period of time like 5 to 10 days
(on account of the limited predictability of turbulent motions
by finite resolution models). Understanding the physical basis
for the long range climate of the planet Earth, or of any planet
for that matter, is a vastly more difficult undertaking because
it would require a very fine knowledge of all mesoscale
(10-100 km) or small scale (10-1000 m) or even molecular processes which do contribute to the energy balance of this enormously
complicated system. The task of quantifying the output of these
processes and establishing their dependence upon the gross
characteristics of atmospheric motions, is known as "parameterization". The first steps in this direction have long been taken
as evidenced by the already striking successes of the most

sophisticated general circulation models at simulating the
Earth present climate. But too many ad hoc assumptions or
semi-empirical inferences are made still, for us to believe that
the present schemes could be extrapolated to widely different
climatic conditions, much less other planetary atmospheres.

For us to appreciate the remarkable complexity of these
problems, let us consider a simplified hydrologic cycle applicable
to atmospheric studies (Figure 21). The atmosphere is indeed a
reservoir of water, in the form of water vapor and condensed
water droplets or ice particles forming clouds. The physical
state in which water is carried along does not materially
affect the direct release or absorption of latent heat, because
clouds form and vanish again in the air mass. Only the latent
heat associated with precipitation of rain or snow is permanently
released to the atmosphere; this is in fact the major process
by which energy is transferred from the sun-heated surface to
the atmosphere. But one does not need to stress the paramount
importance of clouds as radiation screens largely controlling
the amount of reflected solar radiation (albedo) and thus the
solar heat input to the Earth surface. Clouds, together with
other minor gaseous constituants of the atmosphere (water vapor,
CO_2 , ozone ...) also control the amount of infra-red radiation
lost to space. Now, even after a century of ground observation
and more than ten years of space observation, we have yet no
precise quantitative knowledge of the amount, optical thickness
and vertical distribution of clouds in the Earth atmosphere,
much less the knowledge of forecasting when, where and for how
long clouds would form in a planetary atmosphere.

This lack of quantitative knowledge also applies, naturally
to the formation of rain, since major precipitations of rain are
associated with deep convective motions which are very complicated
three-dimensional geophysical flows of their own right : we know
more about the dynamics of the (two-dimensional) general circulation of a planetary atmosphere than about one single deep

convective cell (see for example the presentation by Professor OGURA, in the GARP/JOC Study Conference on the Parameterization of sub-grid scale processes, Leningrad, 1972).

Assume nevertheless that one could, somehow, estimate the amount of precipitation at all grid points of the model. Part of it may be snow (if the lowest atmosphere layer temperature is below the freezing point) falling over land or the polar ice-shelves. Whether this snow cover will remain or not depends upon the temperature and the heat capacity of the ground;but the solar heat input will drop drastically as the albedo of snow is much higher than any rock or vegetation. We have here a typical example of positive feedback between radiation and the water cycle   : if the snow is dry and white, it will absorb little solar energy and remain cold; if, on the other hand, it is wet and gray, it will absorb more and melt faster.

Water may also be stored in the top soil as ground moisture and eventually evaporate back to the atmosphere, at a rate which depends upon eddy transport in the turbulent boundary layer and vegetation. But it is well known that excess water runs off to be carried away to sea. A proper estimate of ground storage and ground hydrology is thus a necessary element of simulating the Earth climate.

Finally, a lot of water precipitates or flows into the sea and evaporates from the sea. Thus, one must be prepared to exert considerable ingenuity for understanding the process of evaporation from the sea surface, a special aspect of the general problem of air-sea interaction, i.e. the dynamic interaction of the turbulent air-flow with the surface waves of the sea. Evaporation is also, to a large extent, controlled by the sea surface temperature which depends in turn upon eddy mixing in the upper sea layer and up-welling associated with large scale oceanic currents. By this effect, the whole oceanic circulation, with a time scale of years, enters the problem of atmosphere dynamics.

We need not say more about convective clouds dynamics, turbulent boundary layer fluxes, radiative balance in the presence of cloudiness, large scale momentum and energy transfer due to orography and the associated lee-waves nor clear-air turbulence : one can see that the outstanding problem of numerical simulation of the Earth atmosphere is the parameterization of all these small scale processes, i.e. the determination of empirical or statistical laws relating momentum, energy, heat or moisture transfers to the general circulation parameters carried in the integration. We need to do this if we are to understand what physical factors determine the climate of our planet as we know it, how the climate would respond to changes of these parameters, how we could act to initiate, or prevent, such climatic modification. I cannot, even briefly, undertake to describe the many research efforts going on presently nor the results obtained so far. I only wanted to convey to those accustomed to the neater problems of fundamental physics or fluid dynamics, an impression of the immensity of the frontier atmospheric physicists are endeavouring to explore,to arrive eventually, at a better and more practical knowledge of our environment.

64

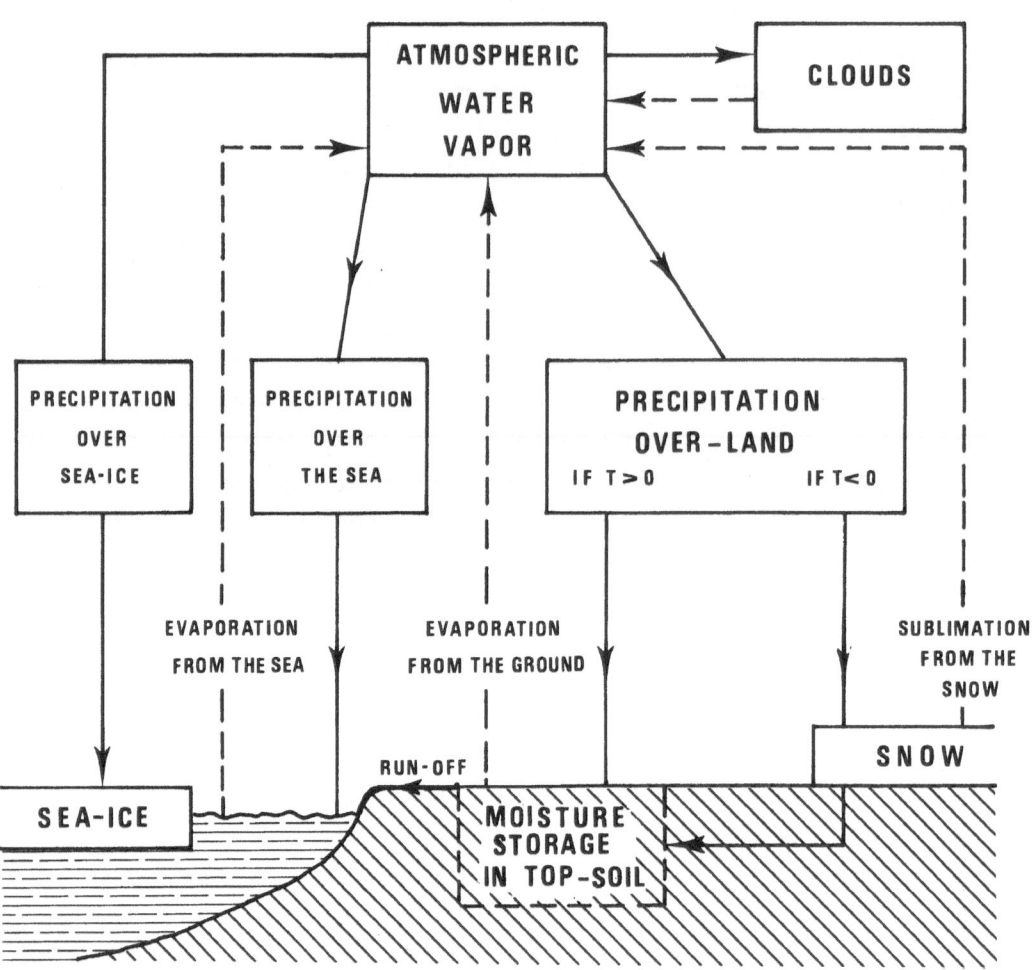

Figure 21
Simplified Hydrologic Cycle

REFERENCES

ARAKAWA,A., Computational Design for Long-Term Numerical Integration of the Equations of Fluid Motion. J. Computational Physics, 1, 119-143 (1966).

BAER,F., and ALYEA,F.N.,Effects of Spectral Truncation on General Circulation Long-Range Prediction. J. Atm. Sciences. 28, 547-480 (1971).

BRYAN,K., and COX,M.D., A numerical investigation of the oceanic general circulation. Tellus. 19, 54-80 (1967).

BRYAN,K., A numerical method for the study of the circulation of the world ocean. J. Computational Physics. 4, 347-376 (1969).

CHARNEY,J.G., Geostrophic Turbulence. J. Atm. Sciences. 28, 1087-1095 (1971).

CROWLEY,W.P., A Global Numerical Ocean Model. J. Computational Physics. 3, 111-147 (1968).

DEARDORFF,J.W., A Three-dimensional numerical investigation of the idealized planetary boundary layer. Geophys. Fluid Dynamics. 1, 377-410 (1970) a.

DEARDORFF,J.W., A numerical study of three-dimensional turbulent channel flow at large Reynolds number. J. Fluid Mech. 41, 453-480 (1970) b.

ELLSAESSER,H.N., Evaluation of Spectral Versus Grid Methods of Hemispheric Numerical Weather Prediction. J. Applied Met. 5, 246-262 (1966).

FISCHER,G., A survey of Finite-Difference Approximation to the Primitive Equations. Monthly Weather Rev. 93, 1-10 (1965).

FOX,D., and LILLY,D., Numerical Simulation of turbulent flows. Reviews of Geophysics. 10, 51-72 (1972).

HOLLAND,W.R., On the wind driven circulation in an ocean with bottom topography. Tellus. 19, 580-600 (1967).

JULIAN, WASHINGTON, HEMBREE and RIDLEY, On the spectral distribution of large scale atmospheric kinetic energy. J. Atm. Sciences. 27, 376-387 (1970).

KAO, and WENDELL, The kinetic energy of the large scale atmospheric motion in wavelength-frequency space. J. Atm. Sciences. 27, 354-375 (1970).

KASAHARA,A., and WASHINGTON,W.M., NCAR global general circulation model of the atmosphere. Monthly Weather Rev. 95, 389-402 (1967).

KASAHARA,A., and WASHINGTON,W.M., General circulation experiments with a six-layer NCAR model, including orography, cloudiness and surface temperature calculations. J. Atm. Sciences. 28, 657-701 (1971)

KURIHARA,Y., On the use of implicit and iterative methods for the time integration of the wave equation. Monthly Weather Rev. 93, 33-46 (1965).

KURIHARA,Y., Numerical integration of the primitive equations on a spherical grid. Monthly Weather Rev. 93, 399-416 (1965).

KURIHARA,Y., and HOLLOWAY,J.L., Numerical integration of a nine-level global primitive equations model formulated by the box method. Monthly Weather Rev. 95, 509-530 (1967)

LEITH,C.E., Atmospheric predictability and two dimensional turbulence. J. Atm. Sciences. 28, 145-161 (1971)

LEOVY,C., and MINTZ,Y., Numerical simulation of the atmospheric circulation and climate of Mars. J. Atm. Sciences. 26, 1167-1190 (1969).

LILLY,D.K., On the numerical simulation of buoyant convection. Tellus. 14, 148-172 (1962).

LILLY,D.K., On the computational stability of numerical solutions of time-dependent non-linear geophysical fluid dynamics problems. Monthly Weather Rev. 93, 11-26 (1965).

LILLY,D.K., The representation of small-scale turbulence in numerical simulation experiments. Proceedings IBM Scientific Symposium on Environmental Sciences. 195-206 (1957).

LORENZ,E.N., Available potential energy and the maintenance of
         the general circulation. Tellus. 7, 157-167 (1955).

LORENZ,E.N., The predictability of a flow which possesses many
         scales of motion. Tellus. 21, 289-307 (1969).

MANABE,S., SMAGORINSKY,J., and STRICKLER,R.F., Simulated climato-
         logy of a general circulation model with a hydrologic
         cycle. Monthly Weather Rev. 93, 769-798 (1965).

MANABE,S., and HUNT,B., Experiment with a stratospheric general
         circulation model. Monthly Weather Rev. 96, 477-539
         (1968).

MANABE,S., and BRYAN,K., Climate and the ocean circulation.
         Monthly Weather Rev. 97, 739-827 (1969).

MANABE,S., SMAGORINSKY,J., HOLLOWAY,J.L., and STONE,H.M.,
         Simulated climatology of a general circulation model
         with a hydrologic cycle. Monthly Weather Rev. 98,
         175-212 (1970).

MANABE,S., HOLLOWAY,J.L., and STONE,H.M., Tropical circulation in
         a Time-Integration of a global model of the atmosphere
         J. Atm. Sciences. 27, 580-613 (1970).

MILES,J.W., Baroclinic Instability of the Zonal wind. Reviews of
         Geophysics. 2, 155-176 (1964).

MIYAKODA, et Al. The effect of horizontal grid resolution in a Atmospheric Circulation Model. J. Atm. Sciences. 28, 481-499 (1971).

MURRAY,F.W., Numerical models of a tropical cumulus cloud with bilateral and axial symmetry. Monthly Weather Rev. 98, 14-28 (1970).

OKLAND,H., Experimental integration of a four-level primitive equation model of the atmosphere. Tellus.21, 359-367 (1969).

OOYAMA,K., Numerical Simulation of the life cycle of Tropical Cyclones. J. Atm. Sciences. 26, 3-40 (1969).

ORSZAG,S.A., Numerical Method for the simulation of turbulence. Phys. Fluids. Supp. II, 12, 250-257 (1969).

ORSZAG, S., Numerical simulation of incompressible flows within simple boundaries. J. Fluid Mech. 49, 75-112 (1971).

PHILLIPS,N.A., The general circulation of the atmosphere : a numerical experiment. Quat. J. Royal Met. Soc. 82, 123-164 (1956).

PHILLIPS,N.A., An example of non-linear computational instability. "The Atmosphere and the Sea in Motion" p. 501-504, Oxford Univ. Press (1959).

RICHTMYER and MORTON, Difference methods for initial value problems. (second edition) p. 404, Wiley, (1967).

ROSENTHAL,S.L., Experiments with a numerical model of tropical cyclone development. <u>Monthly Weather Rev.</u> <u>98</u>, 106-120 (1970) a.

ROSENTHAL,S.L., A circularly symmetric primitive equations model of tropical cyclone development containing an explicit water vapor cycle. <u>Monthly Weather Rev.</u> <u>98</u>, 643-663 (1970) b.

SADOURNY,R., ARAKAWA,A., and MINTZ,Y., Integration of the non-divergent barotropic vorticity equation with an icosahedral-hexagonal grid for the sphere. <u>Monthly Weather Rev.</u> <u>96</u>, 351-356 (1968).

SADOURNY,R., and MOREL,P., A finite difference approximation of the primitive equations for a hexagonal grid on a plane. <u>Monthly Weather Rev.</u> <u>97</u>, 439-445 (1969).

SADOURNY,R., Approximations en différences finies des équations de Navier-Stokes appliquées à un écoulement géophysique. <u>Ann. de Géophysique.</u> (1972, en préparation).

SASAMORI TAKASHI, A numerical study of the Atmospheric circulation on Venus. <u>J. Atm. Sciences.</u> <u>28</u>, 1045-57 (1971).

SHUMAN,F.G., and HOVERMALE,J.B., An operational six-layer primitive equation model. <u>J. Applied Met.</u> <u>7</u>, 525-547 (1968).

SUNDQUIST,H., Numerical simulation of the development of tropical cyclones with a ten-level model. <u>Tellus.</u> <u>22</u>, 359-390 and 504-510 (1970).

WELLCK,R.E., KASAHARA,A., WASHINGTON,W.M., and DE SANTO, G.,
        Effect of horizontal resolution in a finite-difference
        model of the general circulation. Monthly Weather
        Rev. 99, 673-683 (1971).

WIIN-NIELSEN,A., On the annual variation and spectral distribution
        of atmospheric energy. Tellus. 19, 540-559 (1967).

WILLIAMSON,D.L., Integration of the primitive equations over
        a spherical geodesic grid. Monthly Weather Rev. 98,
        512-520 (1970).

METHODS FOR (GENERALLY UNSTEADY) FLOWS WITH

SHOCKS: A BRIEF SURVEY

R. D. Richtmyer

University of Colorado

Boulder, Colorado 80302

I wish to make a few fleeting and very general comments on the
development of computational methods for the dynamics of compressible
(but otherwise completely idealized) fluids. Heat transfer, viscosity, boundary layers, and turbulence are assumed to be absent.

1. Mathematical Formulation of the Initial-Value Problem (IVP)

The flow is described by certain <u>flow functions</u>, such as the
pressure, p, the density, ρ, and the velocity components, u,v,w,
which depend on the space variable, x,y,z, and the time, t. In all
problems except the problem of turbulence (which is not being considered here), these functions are assumed to be piecewise smooth, in
some sense. This establishes a distinction between (1) the smooth
parts of the flow, where the functions are usually assumed to be continuously differentiable as many times as may be desired in any particular connection (for example, for the discussion of the accuracy
of difference equations by Taylor's series expansions of the flow
functions) and (2) discontinuities, such as shocks and interfaces, in
the form of simple jumps across smooth surfaces. Singularities of
lower dimensionality, such as shock intersections and explosion and
implosion centers, are often permitted, but the exact class of admissible functions is seldom stated. The <u>initial data</u> are the functions of x,y,z obtained by setting $t = t_o$ (a constant) in the flow
functions.

The IVP is the problem of determining the flow functions at
later times, when the initial data are given, by use of the conservation laws of mass, momentum, and energy (here understood in their
fundamental or integral form) and perhaps other physical laws. The
equation of state is assumed to be smooth or piecewise smooth.

The consequences of these assumptions include the partial differential equations (PDE) for the smooth parts of the flow and the jump conditions (JC) on shocks, interfaces, and fan boundaries. Conversely, these in turn imply the conservation laws when the flow is piecewise smooth (at least provided there are no singularities of too extreme a nature).

It might be thought that a formulation of the IVP based on these PDE AND JC is complete. However, there are known problems in which a further condition is needed to make the selection unique for given initial conditions, because otherwise the solution may contain either a rarefaction fan or a negative shock or any one of a large family of combinations of rarefaction fans and negative shocks. Hence one must assume the familiar entropy condition, which prevents negative shocks; it can also be expressed as a stability condition for the shock profiles or as a condition on the characteristics.

We should like to know under what conditions (1) a solution exists, and (2) the solution remains piecewise smooth for $t > t_o$ if the initial data are piecewise smooth, (3) the solution is unique, or (4) the solution can be made unique by invoking further physical laws. These questions have been asked before, and I have nothing to contribute to their answers, but I think that those of us who are concerned with the development of computational methods ought to meditate on such questions once in a while, even though our present formulations are probably fully adequate for all "reasonable" problems.

2. Existence, Uniqueness, and General Properties of the Solution of the IVP

In the smooth part of the flow, the existence and uniqueness of the solutions are guaranteed locally by the Cauchy-Kovalevsky theorem on PDE. The theorem says that if sufficiently smooth data (Cauchy or initial data) are given on a sufficiently smooth hypersurface, then a unique smooth solution exists in some neighborhood of that surface (the original proofs were for analytic data on an analytic surface), provided that the surface is nowhere characteristic with respect to the data (or the data with respect to the surface, whichever you prefer). In the fluid-dynamical problem, the hypersurface is the 3-dimensional surface $t = t_o$ in space time, and the characteristic rays (i.e. the bicharacteristics) are the trajectories of the particle and of acoustic signals in space time.

The possible directions of these rays as they pass through the hypersurface $t = t_0$ are determined by the initial data and the equation of state. Since the fluid velocity and the sound speed are finite, the characteristic directions cannot lie in the initial hypersurface (to do so would imply an infinite signal speed); hence the surface $t = t_0$ is nowhere characteristic. Hence the Theorem says that if the initial data are smooth at $t = t_0$ in a 3-dimensional neighborhood of a point x_0 , then there is a unique smooth solution in a 4-dimensional neighborhood of x_0, $t_0)$. This is all that can be said in general, for it is known that shocks can be generated spontaneously within the fluid during the flow.

The simplest example of a problem with nonsmooth initial data is the classical Riemann problem; this is a problem in one space variable x in which the initial functions, u, p, ρ, are constant except for jump discontinuities at a point $x = x_0$. The solution contains a contact discontinuity (usually) and either two shocks, or two rarefaction waves, or one of each, which start at $x = x_0$ and move in opposite directions through the fluid with constant speeds. This is the only "solution" that satisfies the conservation-law jump conditions for shocks, contact discontinuities, and boundaries of rarefaction waves, satisfies the entropy condition for shocks, and satisfies the PDE elsewhere.

Perhaps the next simplest problem is the same as the Riemann problem except that the functions are assumed to be analytic on either side of x_0 , but not necessarily constant. Much work has been done on this problem, and existence and uniqueness have been proved for simplified versions of it (for example, when there is only one function and one PDE), but the general case is still an open question, to the best of my knowledge.

The corresponding multidimensional problem, with initial jumps on a curved surface, is a very open question and is likely to remain so for a long time.

In the absence of existence and uniqueness proofs, one adopts a working hypothesis or conjecture to serve as a basis for the development of computational methods. A reasonable conjecture for the IVP of fluid dynamics would seem to be that if the initial data are piecewise analytic, there exists a unique piecewise analytic solution, at least for some time inverval $t_0 \leq t \leq t_1$ ($t_1 > t_0$). It is assumed that the equation of state is piecewise analytic. [In some problems, "piecewise analytic" can surely be replaced throughout the "piecewise smooth", suitably defined, but not in all cases, for example, not in problems subject to Taylor or Helmholtz instability.] To say that the initial data are piecewise analytic means that space can be divided into cells in each of which the flow functions are analytic for $t = t_0$, the cells are separated by analytic surfaces across which the functions have simple jumps, and the surfaces are bounded by analytic curves (which are the edges of the cells). It is not clear what kind of singularities of the functions or of the surfaces themselves must be permitted at the edges and corners of the cells. (A few examples will be mentioned in Section 5 below.) Hence, the formulation of the conjecture must be left somewhat vague. To say that the solution is piecewise analytic means that space time can be similarly divided into cells, and so on; it doesn't mean of course that the subdivision of space remains for all time the same as for $t = t_0$, for, as already noted, shocks can form spontaneously in the flow; also, shocks and rarefaction fans can diverge from an initial discontinuity, and so on.

In a well-posed IVP, in the sense of Hadamard, a unique solution exists for all initial states of the system defined by some "reasonable" class of admissible functions, and the solution depends continuously, in some sense, on the initial data. Helmholtz instability provides an example of discontinuous dependence on the initial data: A plane surface separates two regions (half spaces), in each of which a fluid is in uniform motion, and slippage (assumed frictionless) occurs on the plane. By introducing a small perturbation of the initial data, in the form of a slight sinusoidal corrugation of the surface and corresponding slight modification of the flow near the surface, a solution can be obtained in which the amplitude of the perturbation increases exponentially with time, as long as the amplitude is small compared to the wavelength. This assumes either incompressibility or that the relative speed of the slippage does not exceed a certain fraction of the sound speed.) Moreover, the exponential growth rate of such a perturbation increases without limit, as the wavelength λ of the corrugations tends to zero. Consequently, given any $\varepsilon > o$ and any $M > o$, one can find a λ small enough so that an initial (very small) perturbation of this wavelength increases by a factor M in a time $\leq \varepsilon$; that is, the solution does not depend continuously on the initial data. It is difficult to make finite-difference methods effective under such circumstances.

If surface tension (or interface tension) exists between the two fluids, then corrugations of wavelength less than some λ_0 are not amplified, and continuous dependence on the initial data is restored, even though longer wavelengths are still unstable.

In the absence of surface tension, the initial surface must be analytic, not merely smooth, for a solution to exist at all. Consider now the case of a piecewise analytic initial surface. A simple case is that of corrugations with edges (alternately convex and concave, as seen from one side) on which plane strips meet, so that a cross section of the surface is a zigzag, ⌇⌇⌇ as in the Figure. The Fourier series method of analyzing the instability breaks down completely in this case even for infinitesimal amplitudes, for although the function that describes the initial surface has a convergent Fourier series, substitution of this series into the equations of motion gives a series which diverges for all $t > t_0$. Nevertheless, I feel that the problem is probably covered by the foregoing conjecture on piecewise analytic problems, and that a solution exists. A study of the corresponding problem with just one edge suggests that a similarity solution may exist, for small $t - t_0$, but the nature of the solution is unknown.

The conclusion of the first two sections of this talk is that, in this general area, it would be highly desirable to know more about the general character of the problems for which we are trying to devise numerical methods.

3. Difference Methods I (Rectangular grid and pseudoviscosity)

Difference methods on a rectangular grid in space-time, with pseudoviscosity treatment of shocks, constitute by far the largest part of the subject, from the point of view of the number and variety of methods and the extent of the theory. In the time available, I cannot give even a survey of this part of the subject, but I wish to make a few remarks, mainly to provide a background for the remainder of my talk.

The first subject I wish to comment on is the question of stability of the difference equations in the smooth part of the flow. Courant, Friedrichs, and Lewy discovered that the now commonest difference scheme for the wave-equation is conditionally unstable, in the sense that the continuous dependence of the solution of the difference equation on the initial data is lost in a rather disastrous way if $c\Delta t$ is $> \Delta x$. When the development of modern computers

in the late nineteen forties made it feasible to apply difference
methods to partial differential equations in practical problems, it
was found that similar phenomena occur in a wide variety of difference schemes for partial differential equations of all kinds. The
von Neumann criterion for stability, which was introduced at that
time, was based on intuitive ideas, but has proved to be highly
reliable, and is still used almost universally in practical calculations. Attempts to put stability theory on a mathematical basis
have led to a vast body of theory. Most of this theory is too
complicated to be useful in practice, but some practically useful
ideas have emerged. One such idea is that of dissipation in a
difference scheme. Formerly, dissipation was avoided whenever
possible, but it is now known that, under certain circumstances, it
can be beneficial rather than harmful. According to a Theorem of
Kreiss, a difference scheme for a hyperbolic system whose degree of
dissipativity is related in a certain way to its degree of accuracy
is certain to be stable under a rather wide set of circumstances.
On the other hand, truly conserved quantities need not be lost. The
Lax-Wendroff equations are dissipative, but mass, momentum, and
energy are exactly conserved, in a certain sense.

I have one comment about the well-known pseudoviscosity method.
The original idea that von Neumann and I had was based on physical
ideas. Shocks were handled at that time by cumbersome shock-fitting
methods, and our idea was to find a simpler and more automatic method.
We were aware, through the work of R. Becker and others, that the
effect of viscosity on a shock is to smear it out so that the transition from the low pressure region to the high pressure one is smooth
rather than discontinuous. If the viscosity coefficient is small, the
transition layer is thin, and the rest of the flow is very nearly
the same as with sharp shocks. Our idea was to write down the differential equations for a slightly viscous fluid, and then solve these
equations by difference methods. With a few minor modifications, such
as the use of a quadratic rather than linear viscosity, the method
turned out to be quite successful in one-dimensional problems.

However, there was a hidden assumption in our work, which was
pointed out only later by P. D. Lax. I wish to mention it, because it
is an assumption that is often overlooked by people who work with
pseudoviscosity methods. Normally, difference methods cannot be expected to give even qualitatively correct results unless the spatial
increment Δx is small compared to the distances in which the

dependent quantities change appreciably; hence, in the present problem Δx ought to be $<< d$, where d is the thickness of the shock.
Instead, we chose $\Delta x \approx \frac{1}{2} d$ to $\frac{1}{4} d$; with this choice, it was unreasonable to expect the method to succeed. Nevertheless, it gave
correct shock speeds and other quantities with an accuracy $\approx 0.1\%$ in
test calculations. Evidently, von Neumann and I were undeservedly
lucky in the particular choice of difference equations we made.

The alternative approach, started by Lax, is to use difference
equations which conserve mass, momentum, and energy exactly. To
avoid severe ringing oscillations behind a shock, the equations
should be suitably dissipative; then the Rankine-Hugoniot jump conditions are satisfied at a shock, because these jump conditions are
based on the conservation laws. Methods that use this idea have been
devised by Lax and Wendroff, by Godunov, and by Rusanov.

Finally, I should like to mention briefly the problem of the
influence of boundaries (external or internal) on the stability of
the difference schemes. In some relatively simple cases, this can
be analyzed by the so-called energy methods. The first attempt at a
more comprehensive theory was the work of Godunov and Ryabenkii.
They gave a stability criterion similar in spirit to the von Neumann
criterion for the smooth part of the flow. It is algebraically much
more complicated than the von Neumann criterion, but may nevertheless
be useful in practical work, owing to the ability of modern computers
to do algebra. Recent work of Bertil Gustafson and H. O. Kreiss have
shown, however, that the Godunov-Ryabenkii criterion can fail in some
cases. Clearly, much more work needs to be done in this area.

4. Difference Methods II (with shock and interface fitting)

In the earliest large scale fluid-dynamical calculations with
shocks, so-called shock-fitting procedures were used (the pseudoviscosity methods were developed only later). These were problems
with just one space variable x , and the method will be described
first for that case.

Let the trajectory of a shock be given by $x = \xi(t)$; then,
$\xi^n \overset{\text{def}}{=} \xi(t^n)$, is the position of the shock at time t^n. The flow
quantities p, ρ, and u have jumps at $x = \xi(t)$. Hence, for each time
t^n, the limiting values $p^{\pm}$, $\rho^{\pm}$, and $u^{\pm}$ of these quantities at
$x = \xi^n \pm o$ are regarded as six further unknowns, in addition to the
values of p, ρ, and u at the regular net points x_j, $j = 1, 2, \cdots$

(generally, ξ^n does not coincide with any of the x_j - see drawing).

With the position ξ^n and the velocity $\dot\xi^n$ of the shock, there are eight new unknowns.

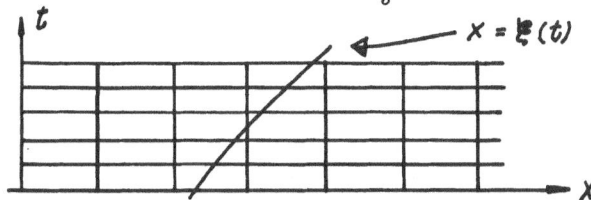

These are governed by the three Rankine-Hugoniot jump conditions and an evident equation such as $\xi^{n+1} = \xi^n + \frac{1}{2} \Delta t \, (\dot\xi^n + \dot\xi^{n+1})$. Four additional equations are written as difference approximations to the partial differential equations involving $p^\pm$, $\rho^\pm$, $u^\pm$, and the values of p, ρ, and u at neighboring net points (this has to be done with considerable care, to achieve stability). The result is a set of simultaneous nonlinear equations, which can be solved by an iterative procedure so as to determine the motion of the shock and to permit the fluid motion on each side to influence the motion on the other side through the jump conditions.

For problems in one space variable containing a single strong shock whose position is known at t = o, shock fitting is highly satisfactory; it gives considerably greater accuracy and resolution than the pseudoviscosity methods.

The need for shock-fitting methods in multidimensional problems is rather urgent, for two reasons. First, considerations of economy generally dictate a coarser net than in one-dimensional problems, so that the pseudoviscosity smearing of the shocks and consequent loss of resolution is more severe. At the same time, the shock configurations are generally more intricate; hence, more, rather than less, resolution is required for their accurate description.

In two-dimensional shock fitting, the position of a shock at time t^n cannot be specified by a single number, but requires the coordinate ξ_ℓ^n, η_ℓ^n ($\ell=1,2,\cdots$) of a number of points sufficiently

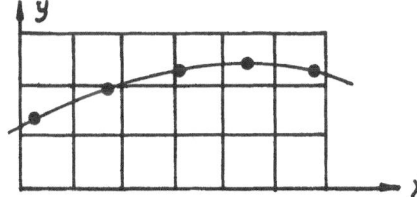

closely spaced on the curve that describes the shock front, as in the figure. The number of new unknowns $p_\ell^\pm$, $\rho_\ell^\pm$ and $u_\ell^\pm$ ($\ell=1,2,\cdots$) is correspondingly increased. New types of instability can arise (e.g. in which the shock front tends to buckle and acquire a saw-toothed irregularity). Much work has been done by various people on the

development of such methods, but it is my feeling that much more
ought to be done.

5. Primarily Analytic Methods; Singularities of Flows

Some problems can be solved by methods that are basically
analytic, such as the Richtmyer-Lewis double-power series Cauchy-
Kovalevski method for the detached shock problem when the shock is
given and one wants to know the flow behind it. The class of problems for which such methods are available is probably rather small,
but analytic methods are of inherently greater accuracy than difference methods. Analytic methods are also often used to provide
starting data for a difference method or to provide auxiliary data
near a difficult singularity.

The success of an analytic or a semianalytic method depends on
knowing the nature of all singularities the flow may possess, and this
is difficult, owing to the nonlinearity of the problem. The triple
shock configuration in

Mach reflection is an example in
two-dimensional flow. Three shock
fronts (one straight and the others
curved) and a slip line (also curved)
meet at a point p, as shown. The
configuration has a similarity property; the distance scale increases
linearly with the time. It would seem natural to assume that the
curves are analytic, and that the flow functions are analytic in each
of the regions I-IV separated by the curves (they are constant in
regions I and II). One might therefore try to represent each curve,
near p, by functions $x(s)$ and $y(s)$, which are power series in s, where
x and y are Cartesian coordinates about p, and s is arc length along
the curve from P, and then to represent the flow quantities in
regions III and IV by double power series in x and y. When this is
tried, one finds that there is no solution having this form near P
that satisfies the differential equations, the various jump conditions,
and the similarity principle. Next, one tries expansions in fractional powers, expansions involving logarithms, etc. These attempts
also fail; hence, the nature of the singularity is not known.

Another singularity that occurs in the same problem (when Mach
reflection is produced in a shock tube) is subsonic compressible

corner flow at the tip of the wedge on the floor of the shock tube
(with nonzero vorticity and variable entropy to boot, but also subject to a similarity principle). It seems likely that this flow
can be expanded in fractional powers of the Cartesian coordinate with
origin at the corner, since this is true of the corresponding incompressible corner flow.

In conclusion, I wish to make a plea for efforts to supplement
numerical work with attempts to understand better the analytic
properties of the flows.

MULTI-LEVEL ADAPTIVE TECHNIQUE (MLAT)

FOR FAST NUMERICAL SOLUTION TO BOUNDARY VALUE PROBLEMS

Achi Brandt

Applied Mathematics Department, Weizman Institute, Rehovot, Israel

In most numerical procedures for solving partial differential equations, the analyst first discretizes the problem, choosing an appropriate operator on a finite-dimensional approximation space, and then devises a numerical process to (nearly) solve the discrete equations. A new technique (MLAT) is proposed, in which discretization and solution processes are intermixed, so as to make both of them much more effective. General boundary-value problems in general domains can be treated by this technique. Similar approach should also be worked out for evolution problems, with the same general aim at automatic codes that eliminate redundancies in both discretization and solution processes.

The main ingredients of MLAT are (a) adaptive discretization and (b) multi-level iterative procedure in which coarser grids constantly participate in solving the equations on finer grids. We shall first describe the technique in general, and then exhibit numerical experiments, including experiments with steady-state Navier-Stokes equations. The advantages of the method, in efficiency, generality and insensitivity, will be discussed. The only disadvantage seems to be the complex programming involved.

1. General description

We write the given differential boundary-value problem in the general form

$$(1) \qquad Au = f$$

where A is a given vector-valued operator whose components represent both the differential equation(s) and the boundary condition(s). f is a given vector-valued function, and u is the unknown function(s), belonging to some space S of functions over a given domain G.

To solve (1) numerically, MLAT uses a sequence of finite-dimensional approximation subspaces $S_1, S_2, \ldots$ S , with <u>local</u> variables and converging accuracy, on which eq. (1) is <u>locally</u> discretized. For instance, one can think of each S_k as a space of net functions, for which FDE (finite difference equations) are formed in the traditional fashion. Or, better still, S_k can be a FEM (finite element method) approximation space of piecewise polynomials for which the discrete problem may be defined as finding that element $u_k \epsilon S_k$ which "best" satisfies (1), i.e., for which $\|Au_k - f\|$ is minimal in some suitable norm. (Cf., e.g., [1].) FEM formulation shares with FDE the property that the coordinates of S_k (the unknowns) can be represented as quantities (i.e., values of the function and, possibly, some of its derivatives) at 'grid-points' of some prescribed grid G_k, and the resulting discrete problem is a system of n_k <u>local</u> algebraic equations

$$(2) \qquad A_k u_k = f_k .$$

That is, each component of (2) is an algebraic equation involving only quantities at some neighboring grid points. f_k and the unknown u_k are n_k-dimensional vectors.

The main idea in MLAT is the simultaneous use of some p grids $G_1, G_2, \ldots, G_p$ (i.e., p approximation spaces $S_1, S_2, \ldots, S_p$. Typically $4 \leq p \leq 7$), chosen successively, each

83

Figure 1

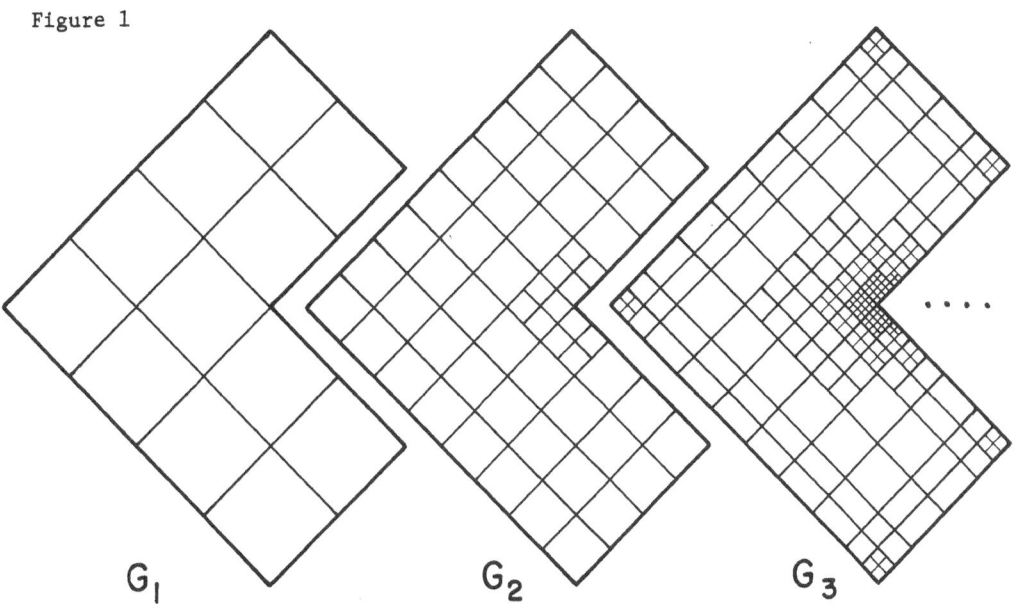

G_1 G_2 G_3 . . . .

G_k much finer than G_{k-1}. (Usually $n_k > 4 n_{k-1}$. See Fig. 1.) The role of the coarser grids, and their associated discrete operators A_j, is threefold:

(a) Each G_k (with its A_k) in its turn is <u>constructed in a manner depending</u> <u>on the approximate solution</u> u_{k-1}, formerly obtained on the coarser grid G_{k-1}: Finer partitions (more grid points) are introduced in regions where u_{k-1} is found to be rough, while higher-order discretizations (e.g., higher-order FDE; or in FEM, higher-degree polynomials) with coarse partitions are used where u_{k-1} exhibits suitable smoothness. Since solutions of elliptic problems are usually very smooth (especially away from boundaries), this adaptive procedure, which can exploit any smoothness wherever it is found, ensures very economic discretizations.

(b) The coarser solution u_{k-1} is also used (interpolated to G_k) as the <u>first</u> <u>approximation</u> $u_k^{(0)}$ for an iterative solution of the newly-formed system (2). See Fig. 2. For the coarsest grid G_1 solutions are easily and inexpensively obtained by conventional methods, since n_1 is small (typically $4 \lesssim n_1 \lesssim 15$).

(c) The first approximation $u_k^{(0)}$ is improved in a sequence of "<u>computational</u> <u>cycles</u>", in which the coarser grids $G_1, \ldots, G_{k-1}$ again play an essential role. Each cycle consists of two processes:

(I) <u>Relaxation sweeps</u>. A relaxation sweep over G_k is a (Seidel-type) process of scanning the grid points (or the grid elements) one by one, correcting, at each point in its turn, the local values of u_k so as to locally satisfy there eq. (2). As a complete method for solution, relaxation converges very slowly (cf., e.g., [2]), but its purpose here is different, namely, only to smooth out the errors. This is effectively achieved in just a few sweeps.

(II) <u>Large-scale corrections through coarser grids</u>. Given an approximate solution $u_k^{(i)}$ to eq. (2), we define for it the "residual problem"

(3) $$A_k^{(i)} v_k^{(i)} \approx \phi_k^{(i)} ,$$

where $\phi_k^{(i)} = f_k - A_k u_k^{(i)} = A_k u_k - A_k u_k^{(i)}$ is the residual, $v_k^{(i)} = u_k - u_k^{(i)}$ is the unknown correction and $A_k^{(i)}$ is some linearization of A_k around $u_k^{(i)}$. For linear A_k we of course take $A_k^{(i)} \equiv A_k$. In nonlinear cases $A_k^{(i)}$ can be taken as the Newton

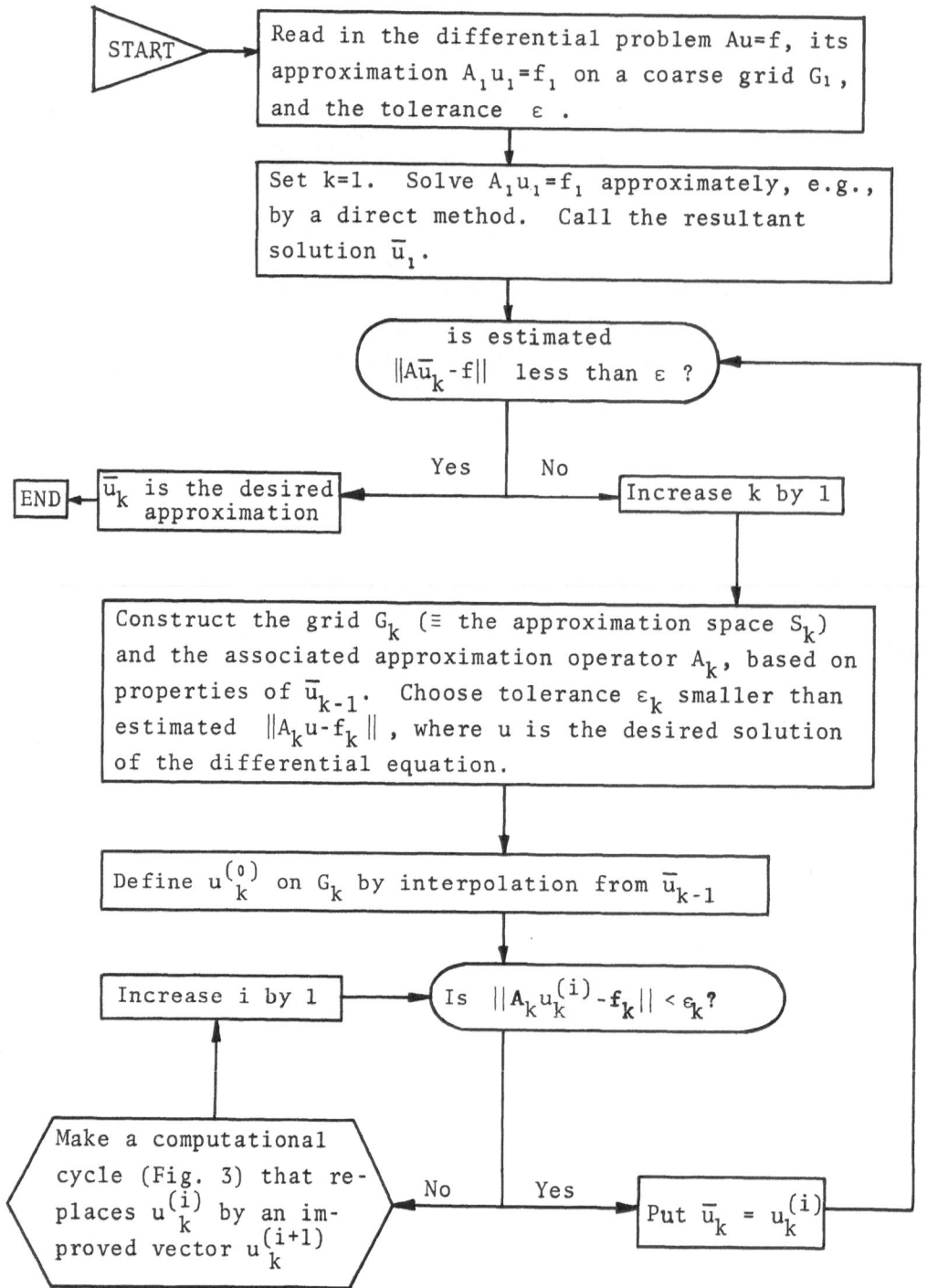

Figure 2. Outline of adaptive schemes

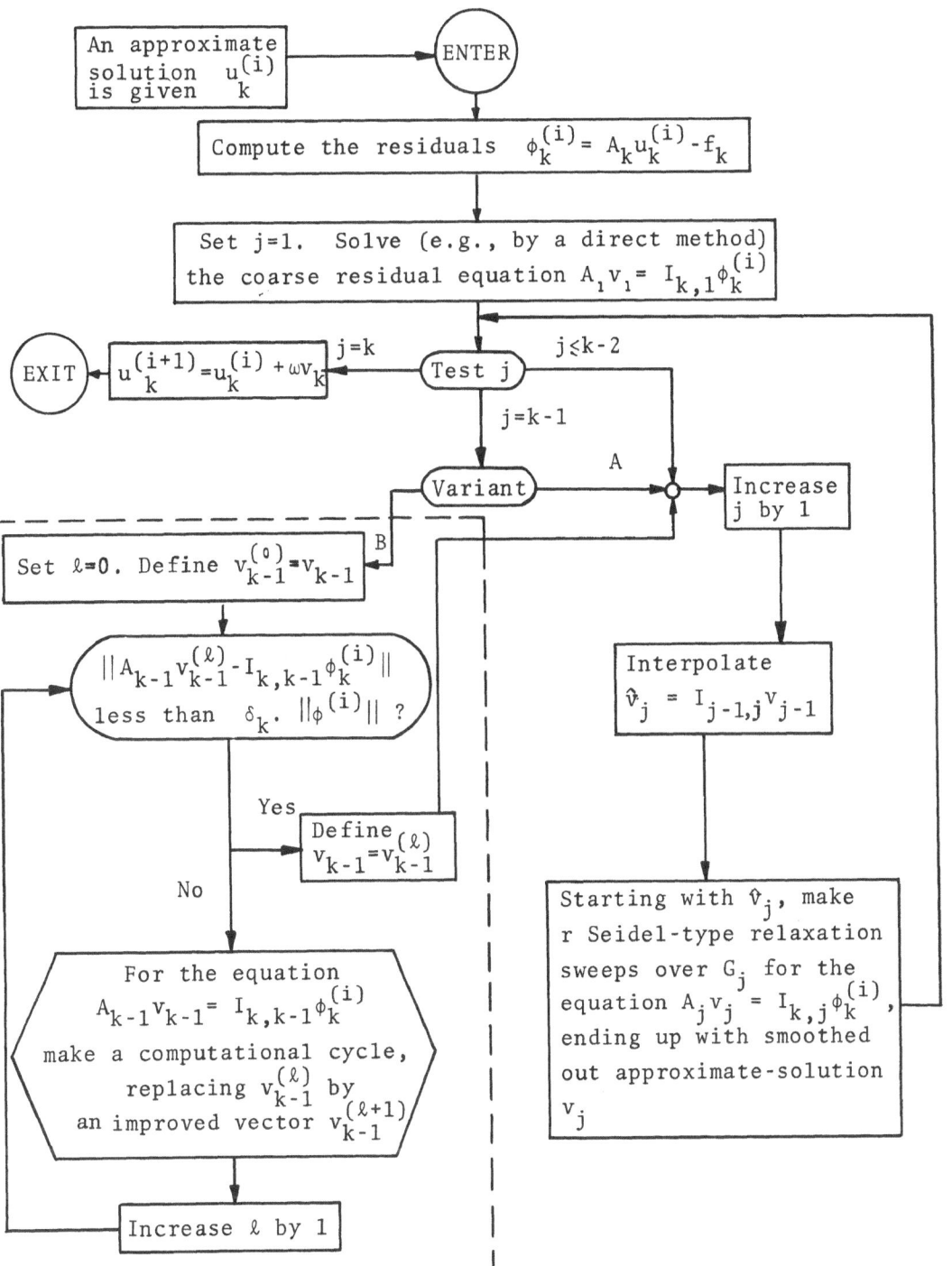

Figure 3. A computational cycle for the equation $A_k u_k = f_k$, replacing an approximate solution $u_k^{(i)}$ by an improved one $u_k^{(i+1)}$. $I_{\ell,j}$ denotes interpolation from G_ℓ to G_j . For nonlinear cases each A_j $(j<k)$ in the flowchart should be replaced by its linearization around $u_k^{(i)}$.

linearization, i.e., the $n_k \times n_k$ matrix whose (μ, ν) component is given by

$$A_k^{(i)}(\mu,\nu) = \frac{\partial\, A_k u_k(\mu)}{\partial\, u_k(\nu)}\,\Bigg|_{u_k = u_k^{(i)}}\ .$$

Once $\phi_k^{(i)}$ is smoothed out (by the relaxations), problem (3) can approximately be solved on coarser grids instead of on G_k, which is of course much easier. The solution of (3), interpolated back to G_k and again smoothed out by relaxations, then serves as a correction to $u_k^{(i)}$. Namely,

(4)
$$u_k^{(i+1)} = u_k^{(i)} + \omega v_k^{(i)}\ ,$$

where ω is a "cycle coefficient", usually taken to be 1. Sometimes, however, $\omega < 1$ is required to help convergence. (See Sec. (2).)

Roughly speaking, the role of relaxation sweeps over any G_j is to liquidate those Fourier components of the error which have wave-lengths comparable to the typical mesh-size of G_j, while corrections through coarser grids liquidates the longer components.

We have used several variants of such computational cycles, two of which are flowcharted in Fig. 3. Variant A is a simplified cycle which we employed most often, with much success. It requires minimal computer storage.

In variant B the idea is to solve (3) approximately using G_{k-1}, where the problem on G_{k-1} is solved by using still coarser grids in a similar fashion. Thus each computational cycle for the k level contains several cycles for the k-1 level, and so forth. This (for FDE formulations) is the original idea of R. P. Fedorenko [3], who also rigorously proved, for 5-point Poisson difference equations on a rectangle, that such an iterative method has rate-of-convergence $\leq q < 1$, where q is independent of the mesh-size. It seems to be the only known method with such a property! This basic contribution, extended by Bakhvalov [4], remained amazingly neglected, at least in the west.

2. Disadvantage of MLAT and existing programs

The main disadvantage of MLAT is the complex programming involved. We have therefore developed so far programs only for some special cases:

(a) Finite-difference solution, on pre-assigned (non-adaptive) sequence of grids, for general second-order, bidimensional equations of the form

$$a_1 u_{xx} + a_2 u_{yy} + a_3 u_x + a_4 u_y + a_5 u = f(x,y)\ , \qquad a_i = a_i(x,y)\ ,$$

in general domains G , with Dirichlet boundary conditions.

(b) Similarly, for the non-homogeneous biharmonic equation $\nabla^4 u = f$ in rectangular domains, with Dirichlet boundary condition.

(c) Similarly, for the Navier-Stokes equation in a square

$$\nabla^4 \psi = R(\psi_y \nabla^2 \psi_x - \psi_x \nabla^2 \psi_y)\ , \qquad\qquad (0 < x,\ y < 1)$$

with boundary conditions $\psi(0,y) = \psi(1,y) = \psi(x,0) = \psi(x,1) = 0$,

$\psi_x(0,y) = \psi_x(1,y) = \psi_y(x,0) = 0$, $\psi_y(x,1) = 1$. We chose for A the crude linearization $A_j^{(i,k)} \equiv \nabla^4$. For large Reynolds numbers R, this linearization is too far off, and to achieve convergence we had to use cycle coefficient $\omega < 1$. (See eq. (4) and Table 2.)

(d) More complicated are programs incorporating the adaptivity feature. Only one-dimensional models have so far been implemented.

Au	G	f	δ	I	K	n_K	r	θ
Δu	[rectangle]	1		1	4	450	1	.63
							2	.59
							3	.58
							4	.57
							5	.58
		$P_2(x,y,e^x,e^y)$		1	2	50	5	.67
				1	3	200	5	.64
				1	4	800	5	.62
				1	5	3200	5	.61
		1		2	4	800	5	.62
				0	4	800	5	.73
$Δ+u_x+u_y$	[circle, triangle]	$P_2(x,y)$		1	5	2000	5-8	.64
								.63
$Δ+u_x+u_y-u$	[triangle]	$P_2(x,y)$		1	5	1500	3-4	.54
							8	.61
$10^m u_{xx}+u_{yy}$, m=1	[rectangle]	1		1	4	450	5-8	.76
m=3								.90
m=5								.90
	[circle]	1		1	5	1200	1	>1.
		$P_3(x,y)$					2	.95
							3	.88
		$5-\dfrac{xy}{\|xy\|}-\dfrac{2y}{\|y\|}$					5	.79
							6-10	.80
$\dfrac{(1,.5)}{(4,.5)}\Big\|\dfrac{(3,2)}{(4,2)}$		$\log(5+xy)+\exp(8+x^2)$					20-30	.83
	[rectangle]	$P_3(x,y)$		1	5	1200	5-10	.81
				1	4	300	5-10	.76
	[heart]	$P_3(x,y)$		1	5	1000	5-10	.80
				2	4	200	5	.67
$\dfrac{(1,.5)}{(2,.5)}\Big\|\dfrac{(3,1000)}{(3,2)}$	[L-shape]	$P_3(x,y)$		1	5	1000	5-10	.80
$\dfrac{(.001,1)}{(.001,1)}\Big\|\dfrac{(1000,1)}{(1000,1)}$	[ellipse]	$P_3(x,y)$		2	4	450	4	.96
	[ellipse]	$P_2(x,y,e^x,e^y)$		2	4	450	3-8	.992
$Δ^2 u$	[rectangle]	1		3	4	961	1-6	>1.
				3	4	961	20	.972
			.25	3	2	49	6	.84
			.25	3	3	225	6	.900
			.25	3	4	961	6	.927
			.25	3	5	3969	6	.938
			.25	3	5	3969	7	.940
			.40	3	4	961	6	.941
			.04	3	4	961	6	.955
		random	.25	3	4	961	6	.93

TABLE 1. Numerical results for linear Dirichlet problems, Au=f in region G with homogeneous boundary conditions. Where diagrams are used for Au, they describe the values of (a_1,a_2) in the four quadrants of the plane, and $Au=a_1 u_{xx}+a_2 u_{yy}$. δ is the $δ_k$ of Fig. 3; when unspecified, variant-A cycles were used. I denotes order of interpolations. P_k denotes k-th order polynomials.

R	K	ω	θ
0-16	2	1.	.85
	3	1.	.915
	4	1.	.932
	5	1.	.934
32	4	1.	.934
48	4	1.	.954
	4	.8	.937
80	4	1.	>1.
	4	.5	.958
160	4	.25	.984
	3	.25	.983
BIG	ANY	$\frac{40}{R}$	$\sim 1 - \frac{20}{R}$

TABLE 2. Convergence rates for the Navier-Stokes equation in a square. With K grids, the finest being a lattice of $2^{k+1} \times 2^{k+1}$ mesh-points ($n_k = 2^{2k+2}$). All computational cycles here were of variant B, with $\delta_k = .2$ and r=6.

3. Numerical Experiments and general conclusions

Tables 1 and 2 summarize some representative results. Our main interest in these experiments was to study θ , the rate-of-convergence of computational cycles, as a function of several problem-dependent and computation-dependent parameters. θ in the tables is actually the asymptotic rate -- its values being usually smaller (i.e., better) in the first couple of cycles. θ is measured as the rate at which the error $e_i = \|A_k u_k^{(i)} - f_k\|$ is reduced, per unit computational work. As unit of work we chose the computational work of one relaxation sweep.

The numerical results may reveal small inconsistencies, owing to parameters which we neglected to control; e.g., the relative positions of the region G and the grids G_i . Nevertheless, several important conclusions seem to emerge:

(a) The rate-of-convergence θ is essentially insensitive to several factors, including the shape of the region G , the right-hand-side f (which influences the convergence only at the first couple of cycles). More significantly, all experiments indicate that $\theta \leq q < 1$, where q is independent of the grid sizes. For Poisson's equations, q ≤ 0.7.

(b) θ does depend on the operator A. Roughly, the closer the operator is to the Laplace operator Δ - the faster is the convergence. This, we believe, is partly due to the interpolation system, which is isotropic. If, instead of interpolating, the equation itself were employed in ascribing values at new grid points, results would be similar to the Δ case in all cases of constant (or moderately varying) coefficients.

(c) Our experiments indicate that the order of interpolation should be one less than the order of the elliptic equation. Lower orders of interpolations cause considerable slower convergence, while higher orders do not make it significantly faster.

(d) A general good value for δ_k appearing in variant B (see Fig. 3) is $\delta_k = .25$.

(e) θ does depend on the number r of relaxation sweeps, but not in a critical way, and good values of r are easily specified. For example, in variant-A cycles for second-order equations, and in variant-B cycles (with $\delta_k = .25$) for fourth-order equations, r = 6 always yields θ very close to the optimum. Another alternative that can be tried is to stop relaxations by some reasonable internal criterion.

(f) For better convergence of the Navier-Stokes solutions with large R, Newton-type linearizations should be used instead of the one described above (Sec. 2).

4. Advantages

The advantage of the adaptive approach in creating efficient discretizations (low number of unknowns n per pre-assigned accuracy ε) is obvious. Also, the technique is non-saturated: The numerical results can be improved indefinitely, by using further grids.

The multi-level technique for solving the discrete equations is the only method that fully exploits the differential origin of these equations, using relations between grids of different sizes. It is not only asymptotically (for very large n) superior to other approaches (as shown in [4]), but even for practical grids it has better convergence rates, as shown in Fig. 4. (The figure is based on Fig. 7 in [5].)

Moreover, the other fast methods shown in the figure lose much of this efficiency when non-rectangular regions are considered and, more significantly, are completely unusable for general non-rectangular grids. Thus, for FEM equations, direct elimination is currently most popular, but this requires $O(n^2)$ operations, as compared with $O(n)$ of MLAT, and is especially at a disadvantage for nonlinear problems. By contrast, MLAT efficiency is not sensitive to the shape of the regions and other parameters. Unlike ADI, Strongly Implicit and other methods, the efficiency here is not critically dependent on optimal choice of certain optimal parameters.

MLAT, as typical to iterative methods, is stable, self-correcting, naturally suitable for nonlinear problems, demands minimal storage, and does not require high computer precision. On the other hand, it is not a purely iterative method, since, at each computational cycle, on the coarsest grid the problem is solved directly. As a result MLAT is convergent also for non-definite problems, like the equation $(\Delta+c)u = f$ for large $c > 0$, where all other iterative methods fail to converge. Solution of such equations is basic for efficient iterative solution of eigen-value problems.

MLAT is very suitable for parallel processing.

Acknowledgement. I would like to thank Y. Shiftan and N. Diner who worked with me on this project.

References

[1] Hubbard (ed.), Numerical Solution of PDE-II, Academic Press 1971. Especially the papers by Babuška; Bramble and Schatz; Marcal; Strang; and references therein.
[2] E. Wachspres, Iterative solutions of Elliptic Systems, Prentice Hall, 1966.
[3] P. P. Fedorenko, Zh. vȳchisl. Mat. Fiz. 4,3 (1964), 559-664.
[4] N. S. Bakhvalov, Zh. vȳchisl. Mat. Fiz. 6,5 (1966), 861-885.
[5] H. L. Stone, SIAM J. Num. Anal., 5 (1968), 530-558.

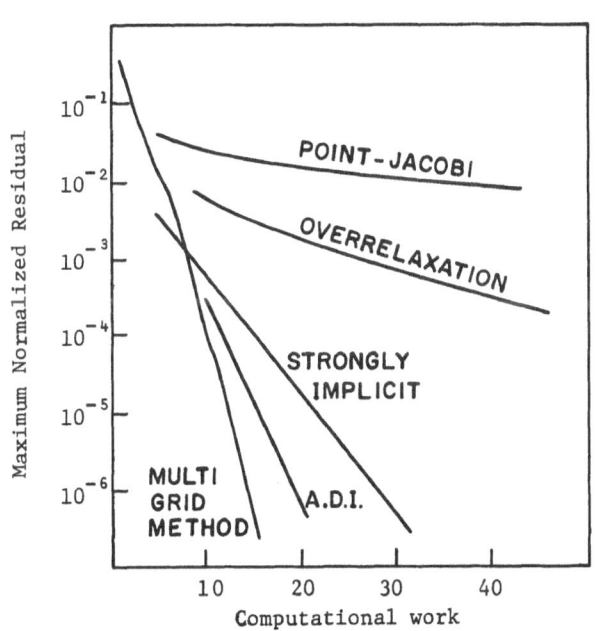

Figure 4. Comparison of computational work required for different iterative methods -- model problem on a 31× 31 grid-point system

A COMPARATIVE STUDY OF DIFFERENCE SCHEMES FOR THE SOLUTION

OF $\dfrac{\partial u_i}{\partial t} + u_j \dfrac{\partial u_i}{\partial x_j} = \nu \dfrac{\partial^2 u_i}{\partial x_j\, \partial x_j} + \dfrac{\partial F}{\partial x_i}$

J.J.D. Domingos, J.M.C. Filipe
Departamento de Engenharia Mecânica
Instituto Superior Técnico
Lisboa, PORTUGAL

0. Introduction

The development of numerical schemes for the solution of Navier-Stokes' equations has been the subject of considerable theoretical and empirical work. However, time-dependent three dimensional flows for arbitrary Reynolds numbers are still beyond the reach of present day computers even if reliable schemes were available, which is not yet granted.

Present work at the authors' group is aimed at the construction of such schemes and the present communication reflects efforts done so far within the framework of general computer programs for coupled sets of quasi-linear parabolic equations in two, three and four independent variables in simple geometries. The general approach has been to test numerical schemes with the model system

$$\frac{\partial u_i}{\partial t} + u_j \frac{\partial u_i}{\partial x_j} = \nu \nabla^2 u_i + \frac{\partial F}{\partial x_i}$$

for which exact solutions have been found for the initial and some initial/boundary value problems.

The model system, which is a generalization of Burger's model to three space dimensions with the inclusion of a source term, as a precise physical meaning as already shown elsewhere [1] , [2] . It possesses the same typical non-linearities and differs from the equations of viscous fluid flow as long as the source term does not coincide with the pressure field (and as such all velocity components at the boundary can not be specified) and the initial velocity field is vorticity free. Related models, where some of the above restrictions are removed have been found, its solution being always reduced to the solution of an initial/boundary value problem for a linear equation [3] .

The existence of exact solutions of a model system which has many of the pecularities of the Navier-Stokes equations is considered a unique advantage, once general time dependent solutions of the N.S. equations and an incomplete understanding of its behaviour makes risky pure empirical approaches and extrapolations.

Using the one-dimensional model system, tests were done on the best known schemes for quasi-linear parabolic equations. These included the explicit, half-implicit * , predictor-corrector and Crank-Nicholson methods has discussed by Lees 4 and Douglas [5] . The conclusions were applied in the construction of a general computer program for coupled equations [6] , [7] , which contains, as an almost obvious extension, the boundary layer equations for fluid flow, including thermal energy transfer and chemical reactions [8] .

The same line of approach was followed for two-dimensions, time dependent

---

(*) called " Modified backward difference scheme " by Lees [4] .

I. Cross derivative scheme

$$u_{xy} \approx [u(x+h,y+h)+u(x-h,y-h)-u(x+h,y-h)-u(x-h,y+h)]/4h^2 \equiv \Delta_{ox}\Delta_{oy}u/h^2$$

$$u_{xx} \approx [u(x+h,y)+u(x-h,y)-2u(x,y)]/h^2 \equiv \delta_x^2 u/h^2 \tag{3}$$

and similarly for u_{yy}. This is the most obvious spatial differencing.

II. Diagonal scheme. Assume $b > 0$. Let α be the $+45^0$ direction from the x-axis, u_α, $u_{\alpha\alpha}$ the directional first and second derivatives in the α direction. Since

$$u_{xy} = u_{\alpha\alpha} - \tfrac{1}{2}u_{xx} - \tfrac{1}{2}u_{yy}$$

we find

$$Lu = (a-b)\, u_{xx} + (c-b)\, u_{yy} + 2b\, u_{\alpha\alpha}$$

$$\approx [(a-b)\,\delta_x^2 u + (c-b)\,\delta_y^2 u + b\delta_\alpha^2 u]/h^2, \tag{4}$$

where $\delta_\alpha^2 u$ denotes $u(x+h,y+h)+u(x-h,y-h) - 2u(x,y)$, and $u_{\alpha\alpha} \approx \delta_\alpha^2 u/(\sqrt{2}h)^2$. If $b < 0$, then the -45^0 direction is chosen as the α direction.

This differencing was studied by Pucci (1958), Bramble and Hubbard (1964), etc. In all these studies, $a-b > 0$ and $c-b > 0$ are assumed in addition to $a,b,c > 0$ and $b^2 - ac < 0$. With small ϵ, while the latter inequalities still obtain, $a-b > 0$ and $c-b > 0$ do not hold simultaneously. Hence we lose the positivity of the solution and further analysis is required.

One could interpret this scheme as an attempt to approximate the diffusion in the ξ direction by that in the two nearest grid directions, and consistency enforces some additional diffusion (which may be negative) in a third direction. This suggests the next approximation.

III. Generalised diagonal scheme. For φ small, $0 \leqslant \varphi \leqslant \beta \equiv \arctan 1/2$, we can generalise II to use the grid points $(x+2h,y+h)$ and $(x-2h,y-h)$. Then

$$Lu = (a-2b)\, u_{xx} + (c-\tfrac{1}{2}b)\, u_{yy} + (\tfrac{5}{2}b)\, u_{\beta\beta}$$

$$\approx [(a-2b)\,\delta_x^2 u + (c-\tfrac{1}{2}b)\,\delta_y^2 u + \tfrac{1}{2}b\delta_\beta^2 u]/h^2. \tag{5}$$

For $\beta \leqslant \varphi \leqslant \pi/4$, we approximate Lu by a combination of $u_{\beta\beta}$, $u_{\alpha\alpha}$, and $u_{\alpha'\alpha'}$, α' being the perpendicular direction to α, and so forth.

For time differencing, we employ fractional step fully implicit schemes of first order accuracy in Δt. With the presence of the cross derivative terms, schemes with higher order accuracy in Δt lead to more complicated implicit equations than the simple and practical tri-diagonal systems that we use. In addition, there are two more fundamental reasons for using fully implicit schemes: (a) The correct decay of a mode of wave number k in one time step is

$e^{-k^2\Delta t} = e^{-(kh)^2\tau}$ where $\tau = \Delta t/h^2$. ($\tau \sim 1$ corresponds to the short time scale, while $\epsilon\tau \sim 1$ corresponds to the long time scale). At very large τ, for long time scale calculations, a partly implicit scheme ($\tfrac{1}{2} \leqslant \theta < 1$) gives a decay per time step $\sim -(1-\theta)/\theta$, while the fully implicit scheme ($\theta = 1$) gives the much better decay $\sim 1/(kh)^2\tau$. (b) Even in a one dimensional problem, only the fully implicit scheme maintains the positivity of the solution at large τ.

For scheme I, we split as follows [Yanenko, 1971]:

The effect of increasing the Reynolds number is clearly appreciated from 4) if the same initial profile is taken; reducing ν(increasing R_e) introduces more and more significant harmonics. Keeping the same number of terms in 4) increasing R_e makes the velocity peak more and more.

R_e=100 was chosen in Fig. 1 for the numerical tests because it gives a typical profile whose spatial resolution at t=0 does not demand too many grid points.

One important practical aspect for very high Reynolds numbers is the restriction put by computer size on the number of grid points and on time step in what concerns computing time.

In fact, if by reducing the time step and the grid size we can always arrive at meaningfull results with all the methods if numerical round-off does not appear significantly, this situation is not realistic because of the practical limitations refered, which become increasingly important in higher dimensions. So, accepting a poor description of the initial profile by lack of grid points, we tested the methods to access the extend to which they smoothed out regions of steep velocity variation without infecting the whole field to such an extend as to become useless.

Some results of the numerical tests are presented in Fig. 2, and detailed in tables I through V. Relative errrors are not shown because they can induce in misinterpretations when the exact values are very small. They can be got from the results given and are always less $\sim$5%, with a typical value lower than 0.1%.

A complete acessement of the methods depends, of course, on the criteria used, and our results show that a best choice can always be a poor one for a particular problem.

For instance, if very small time steps are needed by the description of time evolution and the stability is properly mastered, the explicit method would rank highest regarding both precision and computing time, not mentioning ease of programing. This means that properly used explicit methods for non-linear equations is not such a wrong choice as many people tend to assume.

If general stability, independent of time step, is a priority in acessement, Crank-Nicholson and predictor-corrector methods are favored. As the results show, even with iteration at each time step, which sometimes can reach a high number for satisfying the convergence criteria Crank-Nicholson can be a good choice regarding precision and sometimes computing time. In fact, it shows a good behaviour to deal with steep gradients and the number of iterations reduces significantly when the profiles evolve slowly. This conclusion appeared at first somewhat unexpected when compared with predictor-corrector which does not need iteration. Predictor-corrector, however, is sensitive to the most complex programing and most probably shows a significant dependence on computer structure. However, if Crank-Nicholson shows a fair behaviour regarding computing time, and if many factors points out for its usefulness in solving steady-state problems, it must be considered that the convergence criteria can become a critical factor in its performance. Taking this into account, predictor-corrector appears practically as much more reliable and our results show that in terms of precision it does not significantly differ from Crank-Nicholson. So, predictor-corrector is in many ways the best compromize regarding precision-computing time if core memory is not a limitation.

Half-implicit schemes, which we should more properly call the Lees' scheme seems to have been little noticed and used. It has, however, many advantages which our results probably do not stress enough. In fact, it is unconditionally stable, easy programable and needs only storage at one time level. This is reflected in the computing times shown. Its precision is not too far from Crank-Nicholson and predictor-corrector methods. These strongly recomend it when memory is to be saved which can be the case with simultaneous equations, and almost always when we consider extensions to higher dimensions. This has made it the best choice for solving the boundary layer equations at our group when only a small computer was available 13 and this again pointed out the way to the half-implicit A.D.I. metho-

flows choosing the A.D.I. formulation of Douglas 9 for one equation, extended later on to coupled sets of quasi-linear parabolic equations in two space dimensions [10] , [11] , The success got induced its implementation to coupled sets of three-dimensional time dependent equations [12] .

All the numerical results to be shown for 2D and 3D used the general computer programs written in Fortran IV, and were run on the I.B.M. 360/44 (128K bytes) computer of the " Universidade Técnica de Lisboa".

1. The Model System

The quasi-linear parabolic system

$$1) \quad \frac{\partial u_i}{\partial t} + u_j \frac{\partial u_i}{\partial x_j} = \nu \nabla^2 u_i + \frac{\partial F}{\partial x_i} \qquad\qquad i,j = 1,2,3$$

has the solution 3 ,

$$2) \quad u_i = -2\nu \frac{\partial}{\partial x_i} (\ln \theta)$$

where θ is the solution of

$$3) \quad \frac{\partial \theta}{\partial t} = \nu \nabla^2 \theta - \frac{F}{2\nu}$$

Because there is a unique solution for the initial value problem of 3), and 2) gives a one-to-one correspondance between u_i and θ , 1) possesses also a unique solution (the initial velocity field is vorticity free as needed by the transformation).

A corresponding relation exists for the initial boundary value problem if in 1) only the normal velocity at the boundary is specified.

One interesting fact, closely related with the increased dissipation promoted by the non-linearity in 1), is that to transient solutions of 3), if in the form $\theta = f(t) . \psi (x_i)$ correspond in 1) a steady state.

2. One Dimensional Flow

The model equation, if F=const, is the known Burger's equation.

Applying the transformation 2) and solving 3) for the boundary conditions

$$3) \quad \begin{aligned} u(0) &= 0 \\ u(1) &= U \end{aligned}$$

we get the general solution

$$4) \quad u = 2\nu \frac{\sum\limits_{n=1}^{\infty} a_n \beta_n e^{-\nu \beta_n^2 t} \sin(\beta_n x)}{\sum\limits_{n=1}^{\infty} a_n e^{-\nu \beta_n^2 t} \cos(\beta_n x)}$$

where β_n are the roots of

$$5) \quad \beta_n \, \text{tang} \, \beta_n = \frac{U}{2\nu}$$

and the a_n are to be found from the initial conditions.

From 4), the steady state is, when $t \to \infty$

$$6) \quad u = 2\nu\beta_1 \text{tang} \, \beta_1 x$$

In Fig. 1 the evolution from t=0 to t=∞ is shown for U=100, and a_1=1, a_2=-0.9955, β_1=1.54006, β_2=4.62025

Defining the Reynolds number on the maximum velocity, the Reynolds number is 100 because u_{max} is at x=1.

ds for two and three space dimensions and coupled equations.

These conclusions were expected from known results of theoretical numerical analysis. These are all based on the assumption of a grid size small enough without any specific indication of its smallness. If this is caracterized by a cell Reynolds number, where the typical length is the distance between grid points these conclusions are valid for a cell Reynolds number of one. Increasing the cell Reynolds number by reduction of grid points the experiments have shown that all of the schemes unstable (at least on a computing sense) when the cell Reynolds number increases. The order in which the schemes blow up are the explicit, half--implicit, predictor-corrector, Crank-Nicholson, none of them attaining a cell Reynolds number of ten. This result has considerable practical and theoretical interest and is discussed elsewhere [14] .

3. Two Dimensional Flows

The exact solution was found as described in §1. Making again F=0 and choosing for ease of representation

$$\theta = e^{-2\beta_1^2 \nu t} \cos(\beta_1 x) \cos(\beta_2 y) + \lambda e^{-2\beta_2^2 \nu t} \cos(\beta_2 x) \cos(\beta_2 y)$$

with $\beta_1 = 1.54006$ $\beta_2 = 4.62025$ $\lambda = 1$

Experience acquired with the one-dimensional case with coupled systems of equations of the same type clearly favoured the extension of A.D.I. to coupled equations in two dimensions.

General conclusions would follow the ones already given for the one dimensional case.

4. Three Dimensional Flows

The success and expetience with the extension of A.D.I. to coupled sets in two space dimensions made the development of general computer programs for the three dimensional case an almost obvious choice, which the numerical results got so far validate. Extensive tests have not yet been accomplished due to the demand in computer time.

Of course, in a more significant way than in two dimensions, one can object to A.D.I. its memory requirements because all variables have to be stored at two different time levels. Its accuracy and stability, which can only be accessed when exact solution are available, seems however to more than offset those disadvantages in what regards other proposals.

5. Conclusions

Experience acquired so far shows that for one space dimension and quasi-linear parabolic equations all the methods have their own specified advantages depending on the particular problems and the priorities on stability-accuracy-computing time. For a general method, predictor-corrector represents possibly the best compromise if memory and computing time are not the practical limitations. If so, the half-implicit scheme comes first because of its intrinsic stability and almost the same memory requirements of the explicit scheme.

Extension of the standard schemes of predictor-corrector and half-implicit to coupled sets of quasi-linear parabolic equations prooved successful.

For two and three space dimensions and coupled equations of the quasi-linear parabolic type extension of A.D.I. was found to be practically possible. Limitations in what regards geometry have not been considered.

Most of the conclusions reached using a model system show it to be a meaningfull test. In the absence of exact general solutions of the Navier-Stokes equations it is suggested that the model system used become a standard test to the proposed schemes for the momentum equations. It is argued that this would clear up the unsuitability of certain stabilizing tricks commonly used.

All the conclusions imply a cell Reynolds number of the order of unity. If the cell Reynolds number increases up to ten all of the schemes blow up even those for which a formal proof of stability exists.

Aknowledgements

The cooperation of J.J. Amarante dos Santos of the Divisão de Termodinâmica Aplicada (Instituto Superior Técnico) at all stages of the work and mainly in the extensive tests of A.D.I. is gratefully aknowledged.

To the computing Center of "Universidade Técnica de Lisboa" and specially to Eduardo Martins for the endless hours of testing we extend our best thanks.

This work was financially supported by the NATO Research Grant Nº 438, and Instituto de Alta Cultura Research Project TLE2.

References

1 - J.J.D.Domingos - Considérations sur le potentiel instantané des vitesses dans un écoulement de fluide visqueux, Comptes Rendus Acad. Sciences, Paris 272, 1666, Série A - 1971

2 - J.J.D.Domingos - Mach Number Effects in Turbulent Flows - AGARD Specialists Meeting in Turbulent Shear Flows - London 1971.

3 - J.J.D.Domingos - The Non-linear equation

$$-\frac{\partial \phi}{\partial t} + f(\phi)(\frac{\partial \phi}{\partial x_j})^2 - \nu \frac{\partial^2 \phi}{\partial x_j \partial x_j} = 0 \qquad \text{DTA Report RC/2 - 1971}$$

4 - M.Lees - Approximate solutions of parabolic equations - J.SIAM, 7,2,167-1959

5 - J.Douglas, B.F.Jones - On predictor-corrector methods for non-linear parabolic differential equations - J. SIAM, 11,1,195 - 1963

6 - J.J.D.Domingos, J.M.C.Filipe - EPA 2015 A general computer program for coupled set of quasi-linear parabolic equations in two independent variables - DTA Report RC/11 - 1972

7 - J.M.C. Filipe - Users' guide for EPA2015

8 - J.J.D.Domingos - Numerical solution of the boundary-layer equations for heat, mass and momentum transfer with chemical reaction - DTA Report RC/12

9 - J.Douglas - Alternating direction methods for three space variables. Numer. Math. 4,1,43,1962

10 - J.J.D.Domingos,J.M.C. Filipe - EPA3012 - A general computer program for coupled sets of quasi-linear parabolic equations in three independent variables - DTA Report RC/13 - 1972

11 - J.J.Amarante - Users' guide for EPA3012 - DTA Report RC/14

12 - J.M.C.Filipe - EPA4012 - A general computer program for coupled sets of quasi-linear parabolic equations in four independent variables - DTA Report RC/15 - 1972

13 - J.J.D.Domingos - The numerical solution of the boundary-layer equations and some underlying mathematical principles - DTA Report - 1970

14 - J.J.D.Domingos, J.M.C.Filipe - The practical meaning of theoretical stability in numerical schemes - DTA Report RC/16 - 1972

Appendix

COMPARISON OF ONE DIMENSIONAL SCHEMES

Common basis:

Reynolds number - 100
Number of grid points - 100
Number of digits in the computation - 16 (Double precision IBM 360/44)

Practical steady state $\nu t=0.5$
Computing time + output for IBM 360/44 with 44 PS compiler
For the Crank-Nicholson method: Imposed limit for the number of operations at each time step: 20
Difference between two successive approximations $\varepsilon < 10^{-4}$
For the explicit scheme the practical limit of stability is r=0.5
h=0.01 $r = \Delta t/h^2$

Exact solution: $u = 2\nu \dfrac{\beta_1 e^{-\beta_1^2 \nu t} \sin(\beta_1 x) + \lambda \beta_2 e^{-\beta_2^2 \nu t} \sin(\beta_2 x)}{e^{-\beta_1^2 \nu t} \cos(\beta_1 x) + \lambda e^{-\beta_2^2 \nu t} \cos(\beta_2 x)}$

with $\beta_1 = 1.54006$
$\beta_2 = 4.62025$
$\lambda = -.9955$

Table I

1.1. Behaviour for short times (t=0.0005)

Method	r	Max.absolute error	Computing time -sec
Explicit	0.5	0.783	4.4
Half-implicit	1	5.041	6.3
Crank-Nicholson	1	2.003	9.2
Predictor-corrector	1	2.130	8.9

Table II

Behaviour for intermediate time - t=0.005

Method	r	Max. absolute error	Computing time - sec
Explicit	0.5	0.023	16.7
Half-implicit	1	0.155	24.9
Crank-Nicholson	1	0.149	57.5
Predictor-corrector	1	0.154	51.0

Table III

Behaviour for long time - t=0.01

Method	r	Max. absolute error	Computing time - sec
Explicit	0.5	0.011	29.4
Half-implicit	1	0.068	44.6´
Crank-Nicholson	1	0.055	104.8
Predictor-corrector	1	0.056	96.9

Table IV

Intermediate time - t=0.005

Method	r	Max. Abs. error	Computing time - sec
Half-implicit	10	3.091 (0.155)	6.3 (24.9)
Crank-Nicholson	10	3.051 (0.149)	13 (57.5)
Predictor-corrector	10	1.881 (0.056)	8.9 (51.0)

Values within brackets are for r=1

Table V

Long time - t=0.01

Method	r	Max. Abs. error	Computing time
Half-implicit	10	1.189 (0.068)	15.8 (29.4)
Crank-Nicholson	10	1.061 (0.055)	30.3 (44.6)
Predictor-corrector	10	0.636 (0.056)	20.9 (96.9)

Values within brackets are for r=1

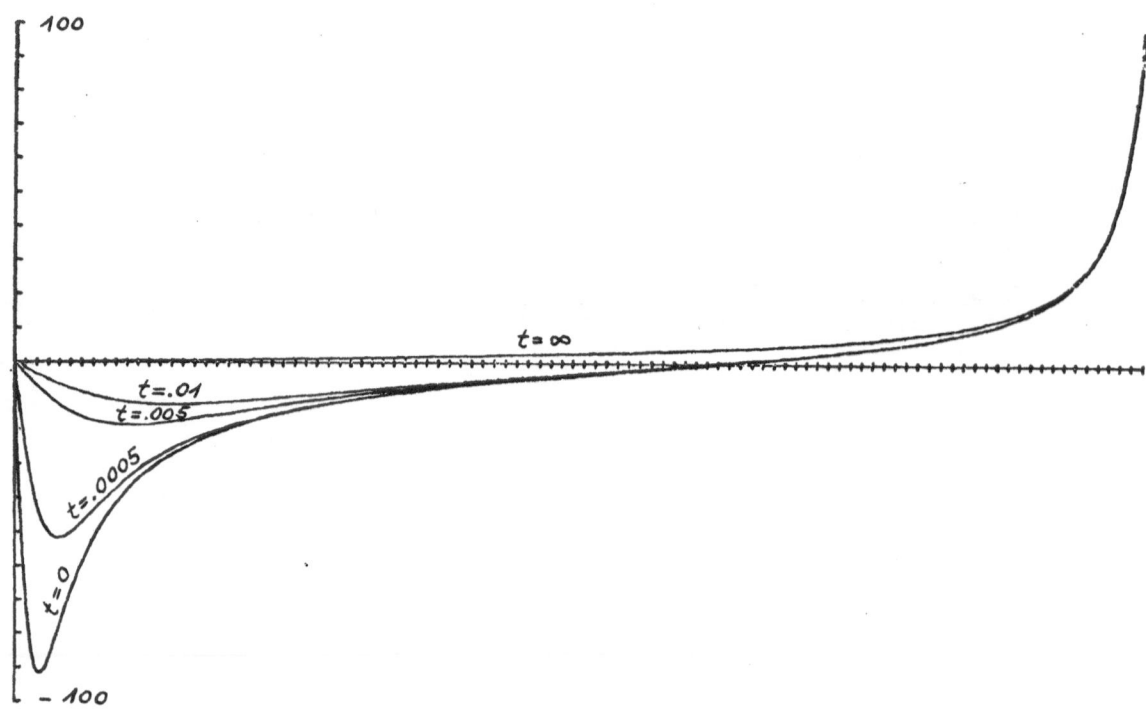

Fig. 1. Analytic Solution

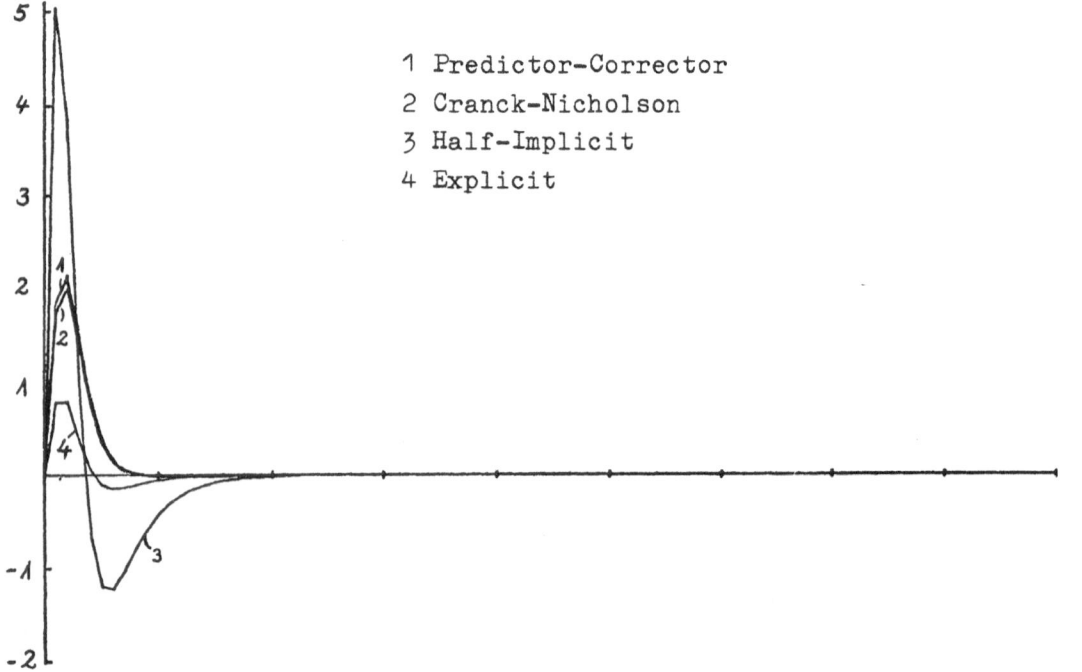

1 Predictor-Corrector
2 Cranck-Nicholson
3 Half-Implicit
4 Explicit

Fig. 2. Absolute Error ($t = .0005$)

APPROXIMATION DES FONCTIONS A DIVERGENCE NULLE

PAR LA METHODE DES ELEMENTS FINIS

par

Michel Fortin*
Université Laval, Québec

1. INTRODUCTION

Nous présentons ici deux approches pour l'approximation des fonctions à divergence nulle, basées sur l'utilisation de la méthode des éléments finis. On sait que les fonctions à divergence nulle jouent un rôle important dans les écoulements des fluides incompressibles ainsi que dans certains problèmes de mécanique des milieux continus. Par ailleurs, la méthode des éléments finis a fait ses preuves dans l'étude des problèmes d'élasticité et de nombreux chercheurs ont tenté de l'appliquer à la mécanique des fluides. Nous voulons résumer ici une étude systématique (Fortin [2]) du traitement de la condition d'incompressibilité.

Nous présenterons aussi pour fixer les idées un algorithme permettant d'appliquer nos résultats au problème de Stokes, mais il devrait être clair que les possibilités d'application sont beaucoup plus vastes.

2. POSITION DU PROBLEME

Etant donné un domaine Ω de $\mathbb{R}^2$, borné de frontière Γ, le problème de Stokes consiste de façon classique à chercher un vecteur $U = \{u, v\}$ et une fonction scalaire p vérifiant:

$$
\begin{aligned}
&- \Delta U + \text{grad } p = F \\
&\text{div } U = 0 \\
&U/\Gamma = 0 \qquad \text{où } F = \{f_1, f_2\} \text{ est connu}
\end{aligned} \qquad (2.1)
$$

On peut formuler de façon plus précise le problème (2.1) dans le cadre fonctionnel suivant: soit

$$V = \{V \mid V \varepsilon \ (H_o^1 \ (\Omega))^2 \text{ et div } V = 0\} \qquad (2.1)$$

On cherche à minimiser sur V la fonctionnelle

$$J(V) = \int_\Omega |\text{grad } V|^2 \ dx - \int_\Omega F \ V \ dx \qquad (2.3)$$

On peut considérer ce problème comme étant une minimisation avec contrainte (div $V = 0$) sur $(H_o^1 \ (\Omega))^2$.

L'approche classique par la méthode des éléments finis est de minimiser $J(V)$ dans un espace de dimension finie V_h dont les éléments sont définis de la façon suivante:

---

* Département de Mathématiques, Université Laval, Québec.

a) On découpe Ω en sous-domaines par exemple des triangles

b) Les fonctions de V_h sont des fonctions continues dont la restriction à chaque triangle est un polynôme de degré R.

c) Les fonctions de V_h sont nulles sur la frontière

d) Les fonctions de V_h seront "à divergence nulle" dans un sens à préciser.

Nous considérerons deux approches pour d)

i) Dans la première approche les fonctions de V_h seront sur chaque triangle des polynômes dont la divergence est nulle partout. On a alors $V_h \subset V$ et nous dirons que l'approximation est interne. Nous allons vérifier au No. 3 qu'il existe de telles approximations.

ii) Dans la seconde approche, nous utiliserons pour V_h des fonctions dont la restriction à chaque triangle est un polynôme dont la divergence sera nulle en moyenne.

On a alors $V_h \subset (H_0^1(\Omega))^2$ mais $V_h \not\subset V$ et nous dirons que l'approximation est externe. Nous verrons que cette approche est possible bien qu'elle entraîne une perte de précision.

3. APPROXIMATION INTERNE

Considérons, pour fixer les idées, un exemple simple: soit un domaine Ω de frontière polygonale décomposé en triangles et considérons une approximation par des fonctions continues, linéaires sur chaque triangle. Un vecteur à divergence nulle sera défini sur un triangle par ses deux composantes.

$$U = \{u, v\}$$
$$u = a_1 x + b_1 y + c_1 \qquad (3.1)$$
$$v = a_2 x + b_2 y + c_2$$

et la condition div $U = 0$ s'exprime par

$$\frac{\partial u}{\partial x} + \frac{\partial v}{\partial y} = a_1 + b_2 = 0 \qquad (3.2)$$

On impose donc une contrainte linéaire sur les coefficients. Ces coefficients sont déterminés de façon classique par les valeurs aux sommets M_1, M_2, M_3 du triangle et la condition (3.2) se traduit à son tour par une relation linéaire entre les valeurs à ces sommets, (où valeurs nodales). On vérifie aisément que le type d'approximation défini ci-haut n'existe pas en général: le nombre de contraintes du type (3.2) est plus grand que le nombre de valeurs nodales. Une telle approximation ne peut exister que pour des triangulations très spéciales où les conditions (3.2) ne sont pas toutes linéairement indépendantes. On est donc conduit à utiliser des polynômes d'ordre plus élevé. Nous nous placerons d'emblée dans un cas favorable (voir [2] pour une discussion générale). Considérons une approximation V_h dont les fonctions sont définies sur chaque triangle par des polynômes d'ordre 4. Un vecteur à divergence nulle est défini par 15 coefficients pour chacune de ses composantes soit en tout trente coefficients. La divergence de ce vecteur est un polynôme d'ordre 3 défini par 10 coefficients que l'on peut déterminer par dix valeurs nodales soit:

1) M_1, M_2, M_3 les sommets du triangle

2) M_4, M_5, M_6, M_7, M_8, M_9 divisant les côtés
   en trois parties égales

3) M_{10} situé au barycentre du triangle

La construction de V_h se fait en deux étapes.

a) On considère un espace V_h formé de vec-
    teurs dont la divergence est nulle au
    bord de chaque triangle (i.e. en M_1 - M_9).
    La divergence est sur le triangle une fonction bulle dé-
    finie par le noeud M_{10}.

On impose ainsi 9 conditions linéaires entre les coefficients
et on peut déterminer les 21 coefficients qui restent en donnant

i) aux sommets du triangle u, v, $\frac{\partial u}{\partial x}$, $\frac{\partial u}{\partial y}$, $\frac{\partial v}{\partial x}$ ($\frac{\partial v}{\partial y} = -\frac{\partial u}{\partial x}$)

ii) aux milieux des côtés u et v

On montre par les méthodes de Strang [4] que si on prend un
vecteur

U = {u, v}, il existe

$U_h \epsilon$ V_h tel que

$||U - U_h|| = O(h^4)$ (3.3)

b) On modifie alors U_h de sorte que

$$\int_{\Gamma i} U_\Gamma \, n_i d \, \Gamma_i = \int_{\Gamma i} U \, n_i d \, \Gamma_i$$ (3.4)

sur chaque côté Γi du triangle.

On montre que U_h peut être choisi de sorte que

$||U_h - U_h|| = O(h^4) \Rightarrow ||U - U_h|| = O(h^4)$ (3.5)

et que la divergence de U_h est nulle

Pour ce faire on modifie uniquement les valeurs de u et v aux
milieux des côtés.

c) Si le vecteur U est la solution du pb. de Stokes (2.1) on
a dans le cas d'un domaine à frontière polygonale et si U est réguli-
ère

$||U - U_h|| = O(h^4)$ (3.6)

Ce type d'approximation n'a pas encore été mis en oeuvre faute
de temps. L'extension au cas tridimensionnel reste à faire.

4. APPROXIMATION EXTERNE

Dans ce cas, on approche V_h de façon un peu semblable à ce qui se passe dans le cas des différences finies, par des fonctions apparentant à un espace plus grand que V, en l'occurence $(H^1_0(\Omega))^2$. Nous remplaçons la condition div $U_h = 0$ par la condition plus faible.

$$\int_T \text{div } U_h \, dT = 0 \text{ pour tout triangle T.} \qquad (4.1)$$

Nous considérerons ici une approximation formée de fonctions dont la restriction aux triangles est un polynôme du second degré.

On peut définir une telle approximation en donnant les valeurs de la fonction:

    i)   aux sommets M_1, M_2, M_3

    ii)   aux milieux des côtés M_4, M_5, M_6

Le principe de la construction est le même que dans le cas précédent: on procède en deux étapes.

A)   On montre par les méthodes de Strang [4] où de Ciarlet et Raviart [1] que si

U = {u, v} ε V il existe un vecteur U_h tel que

$$||U - U_h|| = O(h^2) \qquad (4.2)$$

B)   Le vecteur U_h ne vérifie pas (4.1). On cherche alors un vecteur U_h proche de U_h tel que

$$\int_{\Gamma_i} U_h \, n_i d \, \Gamma_i = \int_{\Gamma_i} U \, n_i d \, \Gamma_i \text{ pour chaque côté } \Gamma_i \text{ du triangle}$$
T. $\qquad (4.3)$

On vérifie que l'on peut choisir U_h tel que

$$||U_h - U_h|| = O(h^2), \text{ en modifiant les valeurs de } U_h \text{ aux milieux}$$
des côtés du triangle. On a alors $\qquad (4.4)$

$$||U_h - U|| = O(h^2) \qquad (4.5)$$

C)   Enfin si le vecteur U est la solution du problème de Stokes (2.1) et U_h la solution approchée dans l'espace V_h ainsi construit, on a

$$||U - U_h|| = O(h) \qquad (4.6)$$

de sorte que l'on a perdu un ordre de précision par rapport à (4.4).

Cela est dû au fait que la condition (4.1) est plus faible que la condition div $U_h = 0$. Pour une méthode voisine voir Oden [3].

5. RESOLUTION NUMERIQUE

Nous avons vu en (2.1) - (2.3) que le problème de Stokes pouvait être compris comme étant un problème de minimisation avec contraintes. Dans la version discrète obtenue par exemple en utilisant l'approximation externe du No. 4, on a sur chaque triangle T une contrainte linéaire et il est facile de définir un multiplicateur de Lagrange permettant de ramener le problème à une minimisation sans contrainte.

Le multiplicateur ainsi défini s'interprète de façon directe comme étant une discrétisation de la pression: la pression discrète est une fonction étagée constante sur chaque triangle.

Nous avons utilisé pour la résolution du problème discret l'algorithme d'Uzawa que l'on peut décrire rapidement ainsi:

i)   On choisit p_h^0 de façon arbitraire

ii) On résoud le problème

$$\Delta_h U_h^n - \text{grad}_h p_h^n = f_h \qquad\qquad (5.1)$$

iii)
$$p_h^{n+1} = p_h^n - \rho \, \text{div}_h U_h^n$$

permet de calculer une pression discrète corrigée.

On démontre que cet algorithme est convergent. Il peut être généralisé à des cas où le problème est non-linéaire.

Les opérateurs div_h, grad_h, Δ_h doivent être définis à partir de la formation variationnelle. (Assemblage).

BIBLIOGRAPHIE

[1]   Ciarlet, P.G. et Raviart, P.A. General Lagrange and Hermite interpolation in   with application to Finite Element Me Method (à paraître).

[2]   Fortin, M. Calcul numérique des écoulements des fluides de Bingham et des fluides mentionnés incompressibles par la méthode des éléments finis. Thèse, Paris 1972.

[3]   Oden, J.T. Key, J.E. On some generalisation of the Stiffness relations for finite deformations of compressible and incompressible finite elements. Nuclear Engineering and Design. 15 (1971) 121-134.

[4]   Strang, G., Fixe, G. An analysis of the Finite Element Method. (à paraître).

A NUMERICAL PROCEDURE FOR A FREE BOUNDARY VALUE
PROBLEM IN THE HODOGRAPH PLANE (*)

Bruno GABUTTI
Laboratorio di Analisi Numerica del CNR - Università - PAVIA - ITALY
Giuseppe GEYMONAT (**)
UERMST Parc Valrose 06 - NICE - FRANCE

1. INTRODUCTION

Let us consider, in the transonic case, the aerodynamic field around thin circular arc airfoils in asymptotically uniform flow with zero incidence. In the study of this problem on the hodograph plane with the direct method (see e.g. [2]) an essential point is the following free boundary value problem.

Problem. Find $(z, c, \mathcal{D})$ such that (see figure 1)

i) $\mathcal{D} = \{(t, \xi) \in \mathbb{R}^2 ; o < t < t_A , \xi_S < \xi < \xi(t)\}$ with $\xi_S \geq 0$, $t_A > 1$, and $\xi(t)$ defined for $t \in]o, t_A[$ a regular monotone increasing function and, if $\xi_S = 0$, tangential to the axis $t = 0$, whose inverse function is $t = t(\xi)$ defined for $\xi \in]\xi_S, \xi_B[$;

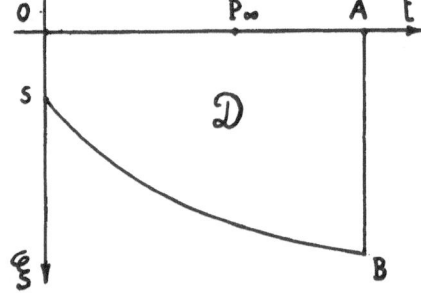

ii) z is regular in $\bar{\mathcal{D}}$ and

$$z_{tt} + t z_{\xi\xi} = 0 \quad \text{on} \quad \mathcal{D}$$

iii) $z = 0$ on $P_\infty A$

iv) $z_\xi = 0$ on OP_∞

v) on BS
   a) $z = - \overset{=}{\psi}$

   b) $c \dfrac{dt/d\xi}{1 + t(dt/d\xi)^2} = -\dfrac{\partial z}{\partial \xi} - \dfrac{\partial \overset{=}{\psi}}{\partial \xi}$ with $c > o$

Fig. 1

vi) on AB $z = z_{AB}(\xi)$ a linear function defined coherently with the conditions iii) and v) a) ;

vii) on OS $\dfrac{\partial z}{\partial t} = c - \dfrac{\partial \overset{=}{\psi}}{\partial t}$ (this condition is absent if $O \underset{=}{} S$)

Here the function $\overset{=}{\psi}$ is defined for $t \geq o$, and for $\xi \geq o$ with $(t, \xi) \neq P_\infty = (1,0)$, where the function is sigular, and corresponds on the

---

(*) Research partially supported by the Laboratorio di Analisi Numerica del CNR of Pavia and by the GNAFA of the CNR.

(**) On leave from Istituto Matematico, Politecnico, TORINO - ITALY.

physical plane to the condition that the asymptotic aerodynamic field is uniform with fixed velocity V_∞ .

About such a problem it is useful to make the following remarks.

1. If the problem has a solution, the properties of $\bar{\bar{\Psi}}$ and the maximum principle imply that $c > o$ and $\dfrac{dt(\xi)}{d\xi} > o$ for $\xi > \xi_S$.

2. Let us define $F(t) = -\dfrac{\partial z}{\partial \xi} (t, \xi(t)) - \dfrac{\partial \bar{\bar{\Psi}}}{\partial \xi} (t, \xi(t))$; if the problem has a solution then the condition v) b gives by integration, for $o \leq t \leq t_A$:

$$\xi(t) - \xi_S = \frac{1}{c} \int_o^t \frac{2\eta \, F(\eta) \, d\eta}{1 \mp (1 - 4\eta \, F^2(\eta)/c^2)^{\frac{1}{2}}}$$

where the sign before the square root is initially negative and is positive for $\eta > t^*$ with t^* such that $4t^* F^2(t^*) = \max 4t \, F^2(t)$; moreover

$$c = 2 \sqrt{t^*} \, F(t^*)$$

3. It is easy to verify that the condition

v) b^{bis}

$$c \, \frac{1}{1 + t(dt/d\xi)^2} = \frac{\partial z}{\partial t} + \frac{\partial \bar{\bar{\Psi}}}{\partial t}$$

is equivalent to v) b. Let us now define $H(\xi) = \dfrac{\partial z}{\partial t} (t(\xi), \xi) + \dfrac{d\bar{\bar{\Psi}}}{\partial \xi} (t(\xi), \xi)$; if the problem has a solution then the condition v) b^{bis} gives by integration

$$t(\xi) = \left(\frac{3}{2} \int_{\xi_S}^{\xi} \left(\frac{c}{H(\eta)} - 1 \right)^{\frac{1}{2}} d\eta \right)^{2/3} ;$$

moreover choosing $t = 0$ in v) b^{bis} one obtains

$$c = H(\xi_S) .$$

4. If there exists a solution it follows from condition v) b^{bis}

$$c = \left(\frac{\partial z}{\partial t} \right) (o, \xi_S) + \left(\frac{\partial \bar{\bar{\Psi}}}{\partial t} \right) (o, \xi_S)$$

and so one can transform condition vii) into the following

vii)^{bis} on OS $\quad \dfrac{\partial z}{\partial t} - \left(\dfrac{\partial z}{\partial t} \right) (0, \xi_S) = \left(\dfrac{\partial \bar{\bar{\Psi}}}{\partial t} \right) (0, \xi_S) - \dfrac{\partial \bar{\bar{\Psi}}}{\partial t} .$

2. THE FIXED BOUNDARY VALUE PROBLEM

Let us suppose that $\mathcal{D}$ and c are <u>fixed</u> and let us consider the problem of determining a z that verifies conditions ii), iii), iv), v) a, vi), vii) . We shall prove that this problem has a unique solution.

Let $V = \{u \in L^2(\mathcal{D}) \; ; \; u_t \in L^2(\mathcal{D}) \;, \; t^{\frac{1}{2}} u_\xi \in L^2(\mathcal{D})\}$ equipped with the natural scalar product

$$((u,v)) = \int_{\mathcal{D}} \{uv + u_t \, v_t + t u_\xi \, v_\xi\} \, dt d\xi$$

and let $\|u\| = ((u, \, u))^{\frac{1}{2}}$ be the corresponding norm. V is a real Hilbert space.

Let $V_0 = \{u \in V \; ; \; u = 0 \text{ on } P_\infty \, A \cup AB \cup BS\}$; with a standard argument one proves that V_0 is closed and that on V_0 the norm

$$\|u\|_0^2 = \int_{\mathcal{D}} (u_t^2 + t u_\xi^2) dt \, d\xi$$

is equivalent to the norm $\|u\|$.

One can prove that $u|_{OS} \in H^{1/3}(0, \, \xi_S)$ and so for every $w \in H^{-1/3}(0, \xi_S)$ the linear form $v \longmapsto L(v) = \langle w, \, v|_{OS}\rangle$ is continuous, where $\langle \, , \, \rangle$ is the pairing between $H^{-1/3}(0, \, \xi_S)$ and $H^{1/3}(0, \, \xi_S)$.

Let $a(u, \, v) = \int_{\mathcal{D}} (u_t \, v_t + t u_\xi \, v_\xi) dt \, d\xi$; it is obvious that $a(u,v)$ is continuous on $V \times V$ and coercive on V_0 .

Now let $g \in V$ be such that $g = 0$ on $P_\infty \, A$, $g = z_{AB}$ on AB and $g = - \bar{\bar{\Psi}}$ on BS and let $\mathcal{U} = \{v \in V \; ; \; v - g \in V_0\}$.

The following result is then a special case of a therem of STAMPACCHIA [4] :

<u>Proposition</u> . <u>For every</u> $w \in H^{-1/3}(0, \, \xi_S)$ <u>there exist a unique</u> $u \in \mathcal{U}$ <u>such that for every</u> $v \in V_0$:

$$(2.1) \qquad\qquad a(u, \, v) = \langle w, \, v|_{OS}\rangle$$

It is easy to verify that (2.1) is a weak formulation of the fixed boundary value problem if one takes $w = c - \dfrac{\partial \bar{\bar{\Psi}}}{\partial t}$.

The nonhomogeneous boundary conditions being regular it is natural to ask whether the solution z inherits the same regularity ; some partial results are given in [2].

3. TWO ALGORITHMS

Starting with a tentative triplet $(z^{(o)}, c^{(o)}, \mathcal{D}^{(o)})$ where $z^{(o)} \equiv 0$, $c^{(o)} = 1$ and $\mathcal{D}^{(o)}$ is defined by a curve $(BS)^{(o)}$ one can generate a sequence $(z^{(k)}, c^{(k)}, \mathcal{D}^{(k)})$ for $k = 1, 2, \ldots$ with two algorithms.

In the first algorithm we shall use the remark 2 of n. 1 and in the second we shall use the remark 3 of n.1.

First algorithm. Step 1. $z^{(k)}$ is the solution of the following fixed boundary value problem :

$$z^{(k)}_{tt} + t\, z^{(k)}_{\xi\xi} = 0 \qquad \text{on } \mathcal{D}^{(k-1)}$$

$$z^{(k)} = 0 \qquad \text{on } P_\infty A$$

$$z^{(k)}_{\xi} = 0 \qquad \text{on } O P_\infty$$

$$z^{(k)} = -\bar{\bar{\Psi}} \qquad \text{on } (BS)^{(k-1)}$$

$$z^{(k)} = z^{(k-1)}_{AB} \qquad \text{on } AB$$

$$z^{(k)}_t - \left(\frac{\partial z^{(k)}}{\partial t}\right)(0, \xi_s) = \left(\frac{\partial \bar{\bar{\Psi}}}{\partial t}\right)(0, \xi_s) - \frac{\partial \bar{\bar{\Psi}}}{\partial t} \qquad \text{on } OS$$

Step 2. Let us define $F^{(o)}(t) = \left[\dfrac{dt/d\xi}{1 + t(dt/d\xi)^2} \right]_{(BS)^{(o)}}$ for $0 \le t \le t_A$

and

$$F^{(k)}(t) = \left[z^{(k)}_{\xi} + \bar{\bar{\Psi}}_{\xi} \right]_{(BS)^{(k-1)}}$$

let

$$c^{(k)} = 2 \sqrt{t^{*(k)}} \; F^{(k)}(t^{*(k)})$$

where $t^{*(k)}$ maximizes $4t[F^{(k)}(t)]^2$

Step 3. $(BS)^{(k)}$, and hence $\mathcal{D}^{(k)}$, is defined as the curve $\xi = \xi^{(k)}(t)$ where

$$\xi^{(k)}(t) = \xi_s + \frac{1}{c^{(k)}} \int_0^t \frac{2\eta\, F^{(k)}(\eta)d\eta}{1 \mp (1 - 4\eta(F^{*(k)}(\eta))^2)^{\frac{1}{2}}}$$

with

$$F^{*(k)}(t) = m \frac{F^{(k)}(t)}{c^{(k)}} + (1-m) \frac{F^{(k-1)}(t)}{c^{(k-1)}}$$

and m is a parameter to be chosen between 0 and 1.

Second algorithm.   Step 1. Identical to step 1 of the first algorithm.

Step 2. Let us define $H^{(0)}(\xi) = \left[\dfrac{1}{1 + t(dt/d\xi)^2}\right]_{(BS)^{(0)}}$ for

$\xi_S \le \xi \le \xi_B$ and

$$H^{(k)}(\xi) = \left[z_t^{(k)} + \bar{\bar{\Psi}}_t\right]_{(BS)^{(k-1)}}$$

let

$$c^{(k)} = H^{(k)}(\xi_S) .$$

Step 3. $(BS)^{(k)}$, and hence $\mathcal{D}^{(k)}$, is defined as the curve $t = t^{(k)}(\xi)$ where

$$t^{(k)}(\xi) = \left[\frac{2}{3}\int_{\xi_S}^{\xi}\left(\frac{1}{H^{*(k)}(\eta)} - 1\right)^{\frac{1}{2}} d\eta\right]^{2/3}$$

with

$$H^{*(k)}(t) = m\,\frac{H^{(k)}(t)}{c^{(k)}} + (1-m)\,\frac{H^{(k-1)}(t)}{c^{(k-1)}}$$

and m is a parameter to be chosen between 0 and 1.

The parameter m characterizes indeed a family of algorithms ; in the numerical experiments we have used m = 0.5 .

4. A CRUDE ERROR ANALYSIS

In practice we have used the procedures obtained by the discretization of each step of the two algorithms.

We shall now describe briefly this discretization (for more details see [2] and [1]) making a crude error analysis.

Step 1. We have used a discretization of local order $O(\Delta^2)$, where Δ is the meshsize, obtained via local Taylor developments.

When $\xi_S = 0$ the matrix of the system so obtained is irreducibly diagonally dominant (in the terminology of [5]) and we have successfully used the Gauss-Seidel iterative method.

When $\xi_S > 0$ the condition vii) gives a matrix reducible and not diagonally dominant ; after eliminating the unknowns corresponding to the boundary OS , one has an irreducible not diagonally dominant matrix. But nevertheless an underrelaxation iterative method converged (even though the convergence became slower when ξ_S increased) with $\omega = 0.5$. (see e.g. [5]).

<u>Step 2</u>. It is well known that to compute derivatives numerically can be an hazardous operation due to roundoff errors. Under suitable conditions one can prove (see e.g. [3]) that there exists an optimal value of Δ that minimizes the sum of absolute values of truncation and roundoff errors ; we have tried to obtain this value of Δ experimentally.

<u>Step 3</u>. The numerical integration has been performed with a Cavalieri-Simpson rule ; here the truncation and roundoff errors are not important.

5. SOME NUMERICAL EXPERIMENTS

Due to the impossibility, at this stage, to give an exact analytic evaluation of the errors described in n.4 , we have adopted the following strategy.

First of all let us remark that all possible types of errors are introduced when we pass from the curve $(BS)^{(k)}$ to the curve $(BS)^{(k+1)}$. So we have fixed a first curve $(BS)^{(0)}$ and we have considered as acceptable these errors if, for $\Delta \longrightarrow 0$, the curves $(BS)^{(1)}$ are numerically convergent.

Moreover, in order to reduce the propagation of the errors through the iterative process, we have carried out a smoothing of the computed curves $(BS)^{(k)}$ with the Gram polynomials of the third order.

We have carried out many computational experiments, both for $\xi_S = 0$ and for $\xi_S > 0$. We have verified that the first and the second algorithm give, for c , opposite-type approximations, at least for $\xi_s = 0$.

In fig. 2 some numerical curves are plotted for different values of $\xi_S = 0$, 0.15 , 0.20 , 0.25 . In fig. 3 there are plotted two sequences of approximations for the value $\xi_S = 0$.

BIBLIOGRAPHY

[1] GABUTTI B., Studio dei profili alari ad arco di cerchio con numero di Mach critico. <u>Atti Acc. Scienze Torino</u>,(1972) 106, pp. 351-369
[2] NOCILLA S., GEYMONAT G., GABUTTI B., Il profilo alare ad arco di cerchio in flusso transonico senza indidenza. <u>Ann. Mat. Pura Appl.</u> LXXXIV (1970) pp. 341-374
[3] RALSTON A., A first course in numerical analysis, McGraw-Hill, New York, 1965
[4] STAMPACCHIA G., Formes bilinéaires coercitives sur les ensembles convexes, <u>C.R. Acad. Sc. Paris</u>, 258 (1964), pp. 4413-4416
[5] VARGA R.S., Matrix iterative analysis, Prentice-Hall, Englewood Cliffs, 1962.

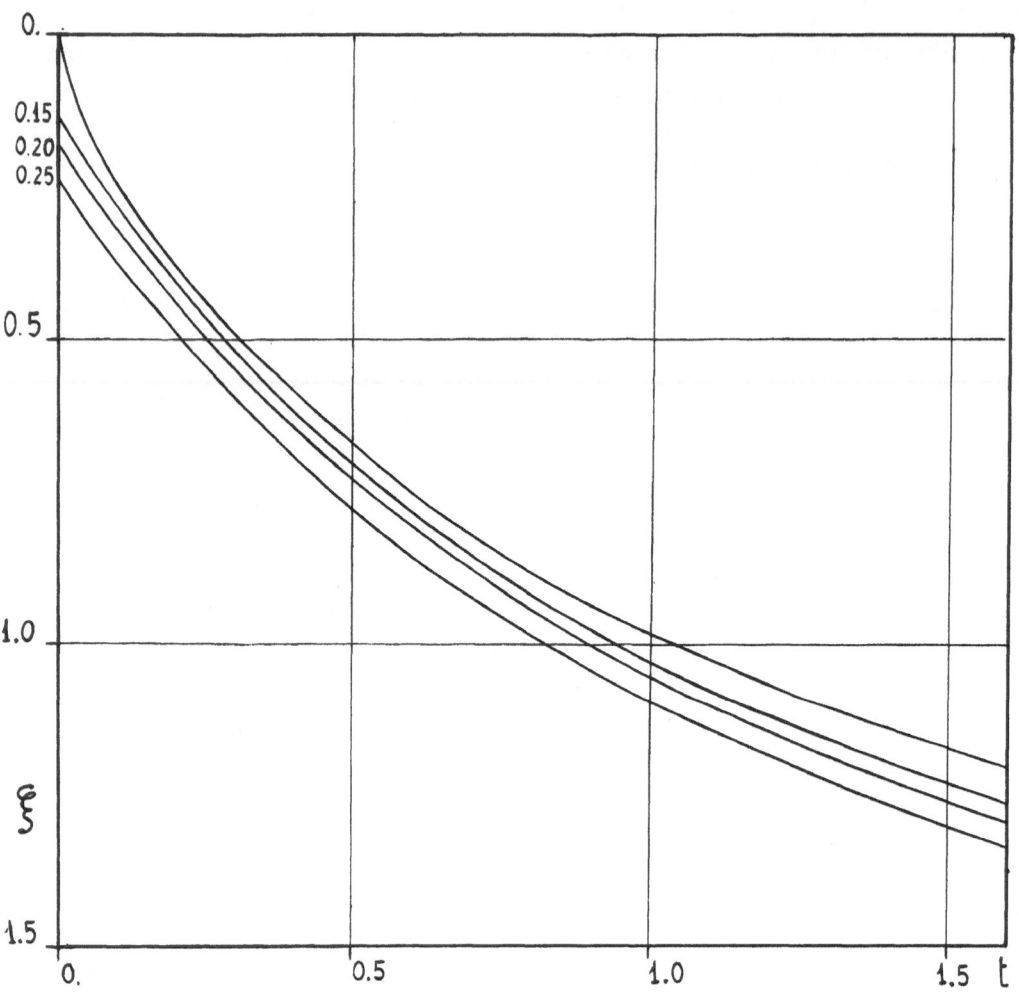

Fig. 2

111

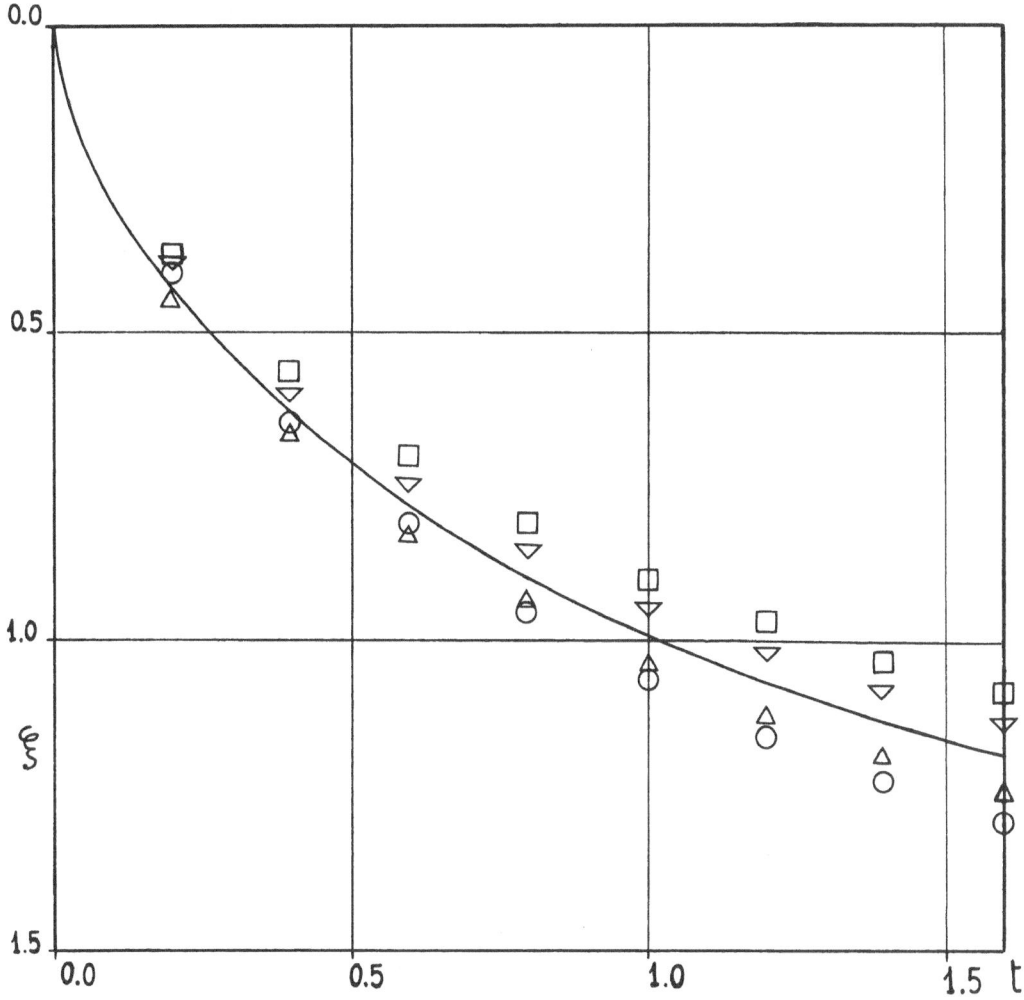

Fig. 3

□ 1ʳˢᵗ iteration 1ʳˢᵗ sequence   ○ 1ʳˢᵗ iteration 2ⁿᵈ SEQUENCE

▽ 2ⁿᵈ iteration   „   „   △ 2ⁿᵈ iteration „   „

The continuous curve is the limit curve after the 7ⁿᵗʰ iteration.

A NUMERICAL STUDY OF A MILDLY NON-LINEAR PARTIAL DIFFERENTIAL EQUATION

L.M. HOCKING
(Department of Mathematics, University College, London)

1. STATEMENT OF THE PROBLEM

When the Reynolds number is greater than some critical value, a plane parallel flow is unstable in the sense that infinitesimal disturbances with a certain range of wave-lengths are amplified. The subsequent history of such disturbances is of considerable interest to the understanding of the transition from laminar to turbulent flow. A theory of the development of an arbitrary initial disturbance at slightly super-critical Reynolds numbers has recently been developed by Stewartson and Stuart(1971), following earlier work by Stuart(1960) and Watson (1960), who considered the development of a modal disturbance. Briefly, the development of the initial disturbance is considered in two stages. In the first stage, for which the time-scale t is that appropriate to the basic parallel flow, the amplification of a narrow range of wave-numbers changes the arbitrary initial disturbance into a modulated wave, which moves downstream with the group velocity, the wave-length being that of the wave which is neutral at the critical Reynolds number. The time scale for the second stage is $\tau = \varepsilon t$, where ε is proportional to the amount by which the Reynolds number exceeds the critical Reynolds number. The modulation of the wave $A(\tau, \xi, \eta)$ is affected by non-linearity at this stage and satisfies the equation

$$\frac{\partial A}{\partial \tau} - a_2 \frac{\partial^2 A}{\partial \xi^2} - b_2 \frac{\partial^2 A}{\partial \eta^2} = \frac{d_1}{d_{1r}} A + k|A|^2 A, \qquad (1.1)$$

where ξ and η are measured in the flow direction and in the transverse direction, respectively. The variation of the disturbance between the bounding walls of the flow is identical to that in the linear stability theory, and the constants a_2, b_2, $d_1 (= d_{1r} + id_{1i})$ and k can be determined for any particular flow from that theory. The solution of (1.1) as $\tau \to o$ must match with the solution of the first stage problem as $t \to \infty$, which gives the initial condition

$$A \sim \frac{\Delta}{\tau} exp\left[-\frac{\xi^2}{4a_2\tau} - \frac{\eta^2}{4b_2\tau} \right], \qquad (1.2)$$

where Δ is arbitrary. The appropriate boundary condition for an initial disturbance which is centred at some point in the flow is

$$|A| \to 0 \quad \text{as} \quad \xi^2 + \eta^2 \to \infty. \qquad (1.3)$$

The equation (1.1) and the conditions (1.2) and (1.3) define the problem.
What is required is the behaviour of the solution as τ increases. Both
analytical and numerical methods were used to answer this question, which turns
out to depend, in a by-no-means obvious way, on the values of the complex constants
in (1.1). The details of the analysis and the main conclusions, together with a
discussion of the implications for the transition problem can be found in Hocking,
Stewartson and Stuart(1972), Hocking and Stewartson(1971,1972). Some details of
the numerical work are explained here, and its important role in the elucidation
of the structure of the solutions exemplified.

2. NUMERICAL PROCEDURE FOR LINE-CENTRED DISTURBANCES

If the initial disturbance has no variation in the span-wise direction, the
equation can be reduced by a suitable scaling to the form

$$\frac{\partial A}{\partial \tau} - \left(1 + i a_i\right) \frac{\partial^2 A}{\partial \xi^2} = A + \left(1 + i \delta_i\right) |A|^2 A, \tag{2.1}$$

with initial and boundary conditions

$$A \sim \frac{\Delta}{\tau^{1/2}} \exp\left[-\frac{\xi^2}{4(1 + i a_i)\tau}\right], \tag{2.2}$$

$$|A| \to 0 \quad \text{as} \quad |\xi| \to \infty. \tag{2.3}$$

The equation was solved using a finite-difference scheme with a Newton
iteration for the non-linear term. With $a_i = \delta_i = 0$ the variables were all real,
but for the other values of the constants, complex arithmetic was used.
(a) Initial Values. Some early calculations were begun by setting $A = 0$
everywhere except at $\xi = 0$, where $A = 1$. This procedure was quite satisfactory,
but involved a considerable amount of uninteresting calculation. The calculation
was usually started by choosing A of the form

$$|A| = \frac{1}{2} \exp\left[-\frac{1}{8}\xi^2\right], \tag{2.4}$$

which was derived from the solution of (2.1) with the term in $|A|^2 A$ omitted,
namely

$$A = \frac{\Delta}{\tau^{1/2}} e^\tau \exp\left[-\frac{\xi^2}{4(1 + i a_i)\tau}\right] \tag{2.5}$$

for suitable values of τ and Δ.

(b) <u>Boundary condition.</u> The outer boundary condition was applied at $\xi = 10$ initially, but the position of this outer boundary was changed whenever the value of $|A|$ close to the boundary exceeded 10^{-5}. An alternative procedure was to use the new variable defined by

$$\bar{\xi} = \xi \left(1 + \xi^2 \right)^{-\frac{1}{2}}, \tag{2.6}$$

with a consequent finite range for $\bar{\xi}$ and a boundary condition

$$|A| = 0 \quad \text{at} \quad |\bar{\xi}| = 1 \tag{2.7}$$

(c) <u>Step lengths.</u> The number of mesh points for the spatial variable was at least 100 and some calculations were done with 400 points. The time step K was decreased as $|A|$ increased, with $K|A|^2$ never exceeding 0.02, a condition suggested by the necessity to balance the terms $\partial A / \partial \tau$ and $|A|^2 A$. Since in some calculations $|A|$ became very large, time steps as small as 10^{-6} were used at times.

3. RESULTS

Two main types of behaviour of the solution were found. For certain parameter values, the value of $|A|$ grew and spread laterally as in the linear solution (2.5) until the non-linear term became important. The value of $|A|$ at $\xi = 0$ then showed a rapid increase in a way that suggested that it was tending to infinity at a finite value of $\tau \left(= \tau_0 \right)$. The relative width of the peak decreased as the height increased. Typical results for the case $a_i = \delta_i = 0$ are shown in Figures 1 and 2.

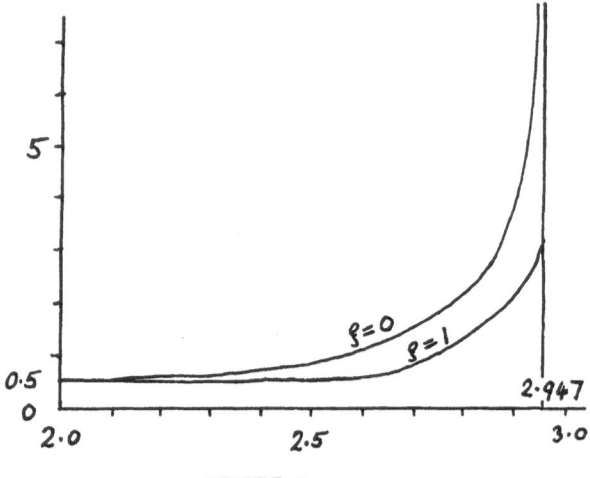

FIGURE 1

The numerical results provided the basis on which the analytical determination of the structure of the bursting solution could proceed. Two such structures were found. In the first, the leading terms in the expansion of $|A|$ for values of τ close to τ_0 , were given by

$$|A| \sim \left[2(\tau_0 - \tau)\right]^{-1/2}\left[1 + \frac{p\, \zeta^2}{(\tau_0 - \tau)\, \ln\,(\tau_0 - \tau)^{-1}}\right]^{-1/2} \qquad (2.9)$$

A comparison of numerical and theoretical values is shown in Figure 3 and the region 1 in Figure 4 gives the parameter values

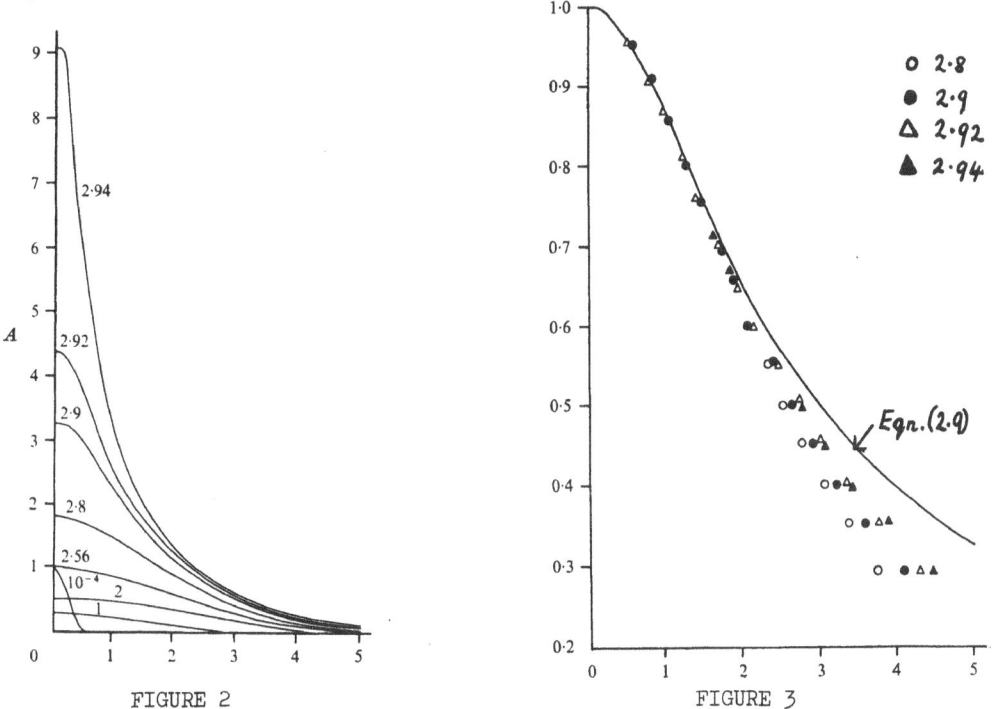

FIGURE 2   FIGURE 3

for which this solution is possible. A second type of burst occurs for parameters in the region 2 of Figure 4. The leading terms are now of the form

$$|A| \sim \left[2(\tau_0 - \tau)\right]^{-1/2} R\left(\zeta(\tau_0 - \tau)^{-1/2}\right), \qquad (3.2)$$

but the approach of the computed solution to this similarity form was very slow (that is, $|A|$ had to become very large). Since accurate numerical solutions when $|A|$ is large require very small step-lengths in ζ in order to obtain an accurate approximation to $\partial^2 A / \partial \zeta^2$, an alternative numerical procedure was adopted. The similarity variables ω, ρ were used, with

$$\omega = |A(\tau, 0)|, \quad \rho = \omega \zeta, \quad A = \omega\, G(\omega, \rho). \qquad (3.3)$$

Some results of calculations using these variables are given in Table 1, for $\delta_i = \sqrt{6}$, $a_i = 0$ and $\tau_0 = 0.42068076$. The values of $F = \frac{d}{dt}\left(\frac{1}{2\omega^2}\right)$

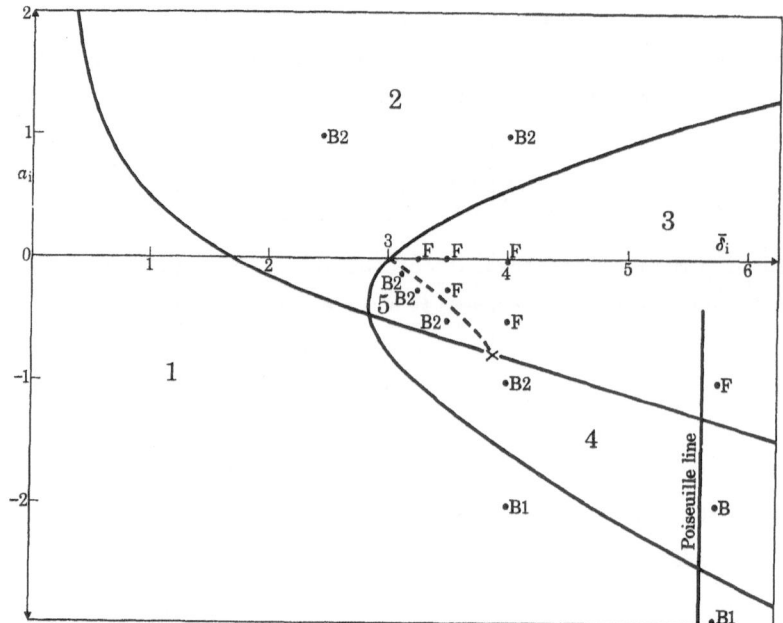

FIGURE 4. Diagram of the regions in the (δ_i, a_i) plane. Bursting solutions occur in regions 1 and 2, finite solutions in regions 3 and regions 4 and 5 are overlap regions. The marked points indicate the form of the solution obtained by numerical computation (B1 and B2 for bursting solutions of the two kinds, F for finite solutions).

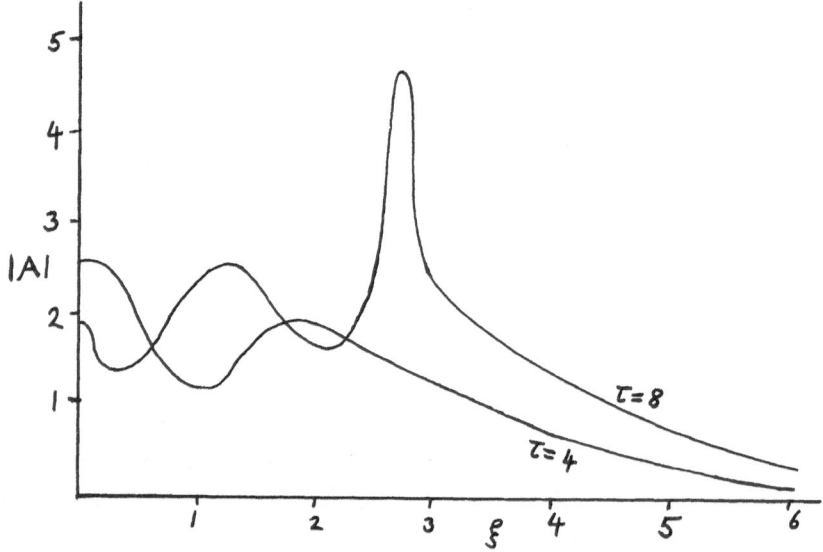

FIGURE 5. Finite Solutions

are given and so is the value obtained from the solution of the quasi-steady similarity equations, to which F should tend as $\omega \to \infty$.

TABLE 1

$\tau_0 - \tau$	ω	F
2.2×10^{-4}	5.25×10^{1}	0.820
5.9×10^{-5}	1.02×10^{2}	0.813
6.5×10^{-7}	0.99×10^{3}	0.791
1×10^{-8}	1.01×10^{4}	0.771
$< 10^{-8}$	1.02×10^{5}	0.754
$< 10^{-8}$	4.35×10^{5}	0.745
-	∞	0.688?

A quite different type of behaviour was found to occur for parameter values corresponding to points in region 3 of Figure 4. The value of $|A|$ at $\xi = 0$ began by increasing but reached a maximum value and thereafter decreased. Subsequent behaviour of the solution exhibited a somewhat random oscillation, with peaks appearing and disappearing at various values of ξ, as shown in Figure 5. Although the solution in some instances was continued to large values of τ, there was no indication of a limiting steady solution being approached, nor even of a regular oscillatory solution. For points in region 3 close to the boundary the first maximum of A was quite large (> 67 for $a_i = 0$, $\delta_i = 3.18$) but as the point moved away from the boundary the first maximum decreased and appeared at increasing values of τ. Analysis showed that a quasi-steady solution of the form

$$A = \lambda L e^{i\omega\tau}\left(\operatorname{sech}\lambda\xi\right)^{1+iM} \tag{3.4}$$

could occur for points in region 3, but the numerical work indicates that this solution is unstable.

There are also two overlap regions 4 and 5 in Figure 4, where more than one type of solution is possible, but the numerical work suggested that whenever a bursting solution is possible, the terminal behaviour of an initial disturbance will be of that bursting type.

4. POINT-CENTRED DISTURBANCES

The solution of (1.1) when the initial disturbance is of the form (1.2), with variations in both ξ and η directions, was also investigated. The additional variable and the additional parameter made the calculation much longer and it was not possible to obtain such detailed results as for line-centred disturbances. Nevertheless, the general pattern was maintained, with bursting solutions of both

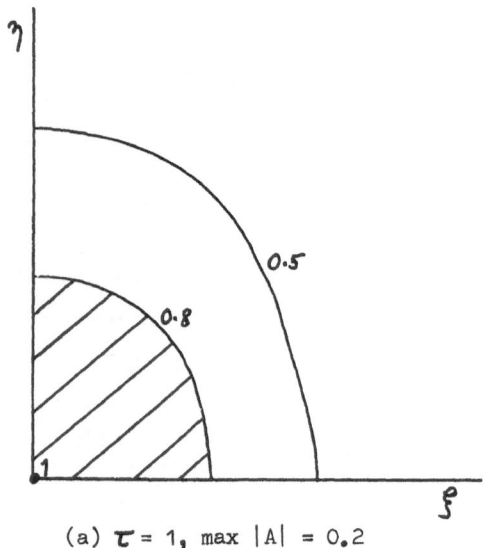

(a) $\tau = 1$, max $|A| = 0.2$

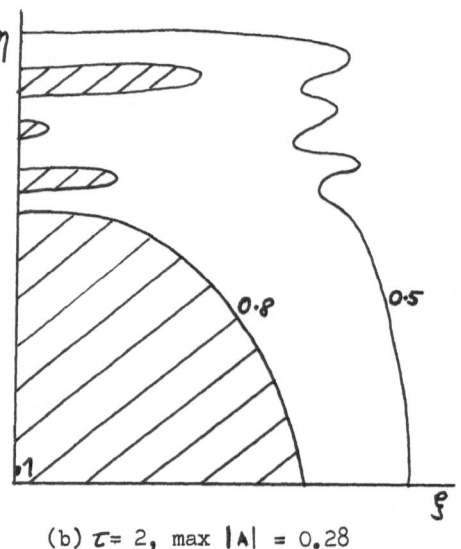

(b) $\tau = 2$, max $|A| = 0.28$

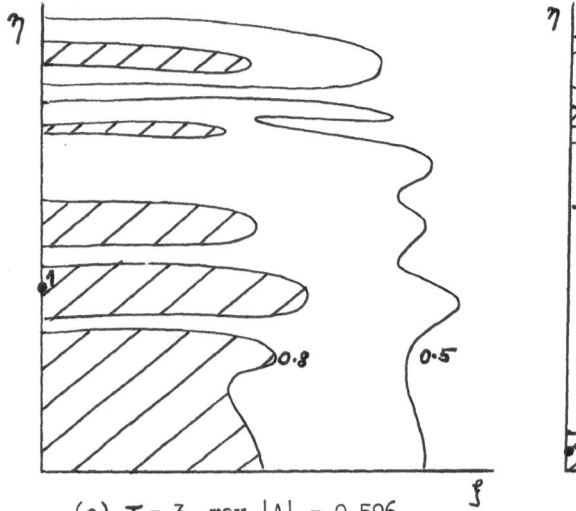

(c) $\tau = 3$, max $|A| = 0.596$

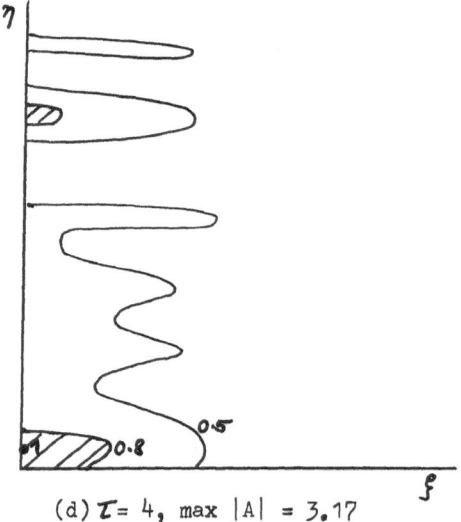

(d) $\tau = 4$, max $|A| = 3.17$

FIGURE 6. Contours of $|A| = $ constant in the (ξ, η) plane

kinds and finite solutions occurring for different ranges of the parameters.

The numerical procedure was similar to that described above. An alternating direction method was used for the two spatial derivatives, with different forms for the non-linear terms at each half-stage of each iteration.

The chief point of interest was to determine the behaviour of a point-centred disturbance when the parameters had the values corresponding to plane Poiseuille flow between fixed walls. The numerical results showed that such a disturbance always remained finite, with local peaks appearing at various points in an apparently disordered fashion. Four stages in the solution are shown in Figure 6 where the ξ and η variables have been scaled so that the initial value, as given by (1.2) has circular contours of $|A|$.

5. CONCLUSIONS

There are two general conclusions which can be drawn from the numerical investigations reported in this paper.

The equation solved has a very mild non-linearity and yet the solutions exhibit a remarkable variety of possible forms. This fact should perhaps stand as a warning against drawing conclusions from limited amounts of numerical evidence when non-linear equations of the complexity of the Navier-Stokes equations are being solved.

In arriving at conclusions about the structure of the solutions of this equation, both analysis and numerical work were needed and the investigation proceeded with a continuous interchange between the two methods of examining the equation. The success of such cooperation in this instance is evidence that more may be achieved by a combination of analytical and numerical techniques than can be obtained by either in isolation.

REFERENCES

Hocking,L.M., Stewartson,K. and Stuart, J.T.   J.Fluid Mech.51, 705-735 (1972)

Hocking,L.M. and Stewartson,K.   Mathematika 18, 219-239 (1971)

Hocking,L.M. and Stewartson,K.   Proc.Roy.Soc.A.   326, 289-313 (1972)

Stewartson,K. and Stuart,J.T.   J.Fluid Mech. 48, 529-545 (1971)

Stuart,J.T.   J.Fluid Mech. 9, 353-370 (1960)

Watson,J.   J.Fluid Mech. 9, 371-389 (1960)

NUMERICAL EXPERIMENTS WITH THE COMPRESSIBLE
NAVIER-STOKES EQUATIONS

Clarence W. Kitchens, Jr.

Ballistic Research Laboratories
U.S. Army Aberdeen Research and Development Center
Aberdeen Proving Ground, Maryland

INTRODUCTION

Several numerical solutions of the Navier-Stokes equations have been obtained for compressible viscous flows, such as those of Crocco [1], Thommen [2], Scala and Gordon [3], Allen and Cheng [4], and Trulio and Walitt [5]. The flow detail in these numerical solutions is quite remarkable, but in certain regions of the flow field the results may be inaccurate. Since there are no rigorous proofs of stability or convergence of the numerical methods employed for these solutions nor any proofs of uniqueness, we should determine the accuracy of any such numerical computation by comparing it against known analytical solutions and experimental measurements.

This paper describes a series of numerical experiments carried out with a set of finite-difference equations developed by Scala and Gordon [6]. Finite-difference approximations of the time-dependent Navier-Stokes equations are used to calculate the two-dimensional flow field around a transverse circular cylinder accelerated to a supersonic velocity. The validity of our numerical results is determined by comparison with analytical bow shock solutions and experimental wake measurements.

Numerical results are presented for M_∞ = 2.0 flow past circular cylinders at freestream Reynolds numbers Re_d of 15.0, 46.8, 157.2 and 704.6, based on cylinder diameter. The numerical solutions are presented in terms of pressure, density, temperature and velocity profiles in the bow shock and wake regions. Experimental wake measurements are presented for low Reynolds number flow at M_∞ = 2.0. These measurements are compared with the numerical computations for Re_d = 157.2.

FORMULATION OF THE NUMERICAL MODEL

The Navier-Stokes equations can be expressed in Cartesian tensor notation as

$$\rho \frac{Du_i}{Dt} = \frac{\partial \tau_{ij}}{\partial x_j} \tag{1}$$

with the stress tensor τ_{ij} given by

$$\tau_{ij} = (- p + \lambda \frac{\partial u_m}{\partial x_m}) \delta_{ij} + \mu (\frac{\partial u_i}{\partial x_j} + \frac{\partial u_j}{\partial x_i}) ; \tag{2}$$

where ρ is the density, D denotes the comoving derivative, u_i are the velocity components corresponding to the x_i Cartesian coordinates, t is the time, p is the pressure, λ and μ are the viscosity coefficients, and δ_{ij} is the Kronecker delta.

The viscosity coefficients in (2) are assumed to be related by Stokes postulate

$$3\lambda + 2\mu = 0.\tag{3}$$

We also use the continuity equation

$$\frac{D\rho}{Dt} + \rho\,\frac{\partial u_i}{\partial x_i} = 0,\tag{4}$$

the energy equation

$$\rho c_v\,\frac{DT}{Dt} + p\,\frac{\partial u_i}{\partial x_i} = \frac{\partial}{\partial x_i}\left(k\,\frac{\partial T}{\partial x_i}\right) + \lambda\left(\frac{\partial u_i}{\partial x_i}\right)^2 + \mu\,\frac{\partial u_i}{\partial x_m}\left(\frac{\partial u_i}{\partial x_m} + \frac{\partial u_m}{\partial x_i}\right),\tag{5}$$

and the equation of state for a perfect gas

$$p = \rho\overline{R}T;\tag{6}$$

where c_v is the specific heat at constant volume, T is the temperature, k is the thermal conductivity, and $\overline{R}$ is the gas constant.

The time dependence in equations (1), (4) and (5) is retained and we seek steady-state solutions of the governing equations by time-wise integration from a prescribed initial state. The steady-state solution is thus approached asymptotically with time. The set of difference equations used to approximate the governing equations was developed by Scala and Gordon [6]. The set of difference equations is formulated by splitting the full system of equations into hyperbolic subsystems and parabolic terms which are differenced separately using an alternating explicit-implicit scheme developed by Gordon [7]. These subsystems are then combined to form the final equations.

We treat time-dependent flow around a circular cylinder using cylindrical coordinates (R,θ) with θ measured from the forward stagnation point. Figure 1 shows a spark shadowgraph of the flow field past a transverse cylinder at Mach two and $Re_d = 15{,}900$. The bow shock, wake shock and inner wake are all visible in this shadowgraph. Separation occurs at this Reynolds number and a separation shock and wake neck are also present in the flow field. The treatment of the time-dependent Navier-Stokes equations requires the specification of initial conditions (t = 0) throughout the flow field and boundary conditions (t ≥ 0) on appropriate boundaries. For times ≤ 0 in our computations we assume a uniform flow past a porous cylinder with freestream conditions throughout the flow field.

For times t > 0 the velocity components inside the cylinder and on its surface are reduced from their initial freestream values to zero. The following boundary conditions are specified for the velocity components on the cylinder

$$U(\theta_i) = \beta U_\infty(\theta_i)$$

and

$$\tag{7}$$

$$V(\theta_i) = \beta V_\infty(\theta_i);$$

where U and V are the radial and tangential velocity components, the subscript (∞) denotes freestream conditions, and

$$\beta = \begin{cases} (t_1 - t)/t_1 & \text{for} \quad 0 \leqslant t \leqslant t_1, \\ 0 & \text{for} \quad t > t_1. \end{cases} \tag{8}$$

Physically, we treat flow past a porous cylinder with the amount of mass flux through the surface decreasing to zero between times $t = 0$ and $t = t_1$. The no-slip condition is specified on the surface for times $t \geqslant t_1$. We specify the temperature boundary condition on the cylinder by assuming an adiabatic wall

$$\left(\frac{\partial T}{\partial R} \right)_{R = R_o} = 0, \tag{9}$$

where R_o is the cylinder radius.

The specification of boundary conditions far from the cylinder poses a difficult numerical problem. We have used two techniques in this study in an attempt to determine the influence of the downstream boundary conditions on the accuracy of the numerical solution. In the first method we specify freestream velocity components and temperature along an outer boundary located at a finite distance from the cylinder. Following Scala and Gordon [3] freestream density values were specified along the finite outer boundary whenever the flow velocity at a mesh point on this boundary was directed into the computation region.

In the second method, suggested by Sills [8], we choose a coordinate transformation to map the interval $[0,\infty]$ of R onto a finite interval $[0,1]$. We transform the radial coordinate by the equation

$$\xi = 1 - e^{-\alpha (R-R_o)}. \tag{10}$$

The arbitrary constant α is a stretching factor which determines the location of the image points. This transformation permits the specification of boundary conditions at infinity while allowing the interior grid points to be concentrated near the cylinder. It should be noted that the use of this transformation results in a solution which is inherently asymptotic at the outer boundary. As long as the derivative $\partial/\partial\xi$ remains bounded as $\xi \to 1$ in the computational plane, then $\partial/\partial R \to 0$ as $R \to \infty$.

We make a symmetry assumption about the flow field to reduce the computer storage requirements and computation time. We specify symmetry boundary conditions

$$\frac{\partial U}{\partial \theta} = 0, \quad \frac{\partial T}{\partial \theta} = 0, \text{ and } V = 0 \tag{11}$$

along boundaries at $\theta = 0°$ and $\theta = 180°$. For the calculations using transformation (10) we additionally specified a symmetry condition for the density along these boundaries. Figure 2 shows the domain of computation and boundary conditions used for two cases with a finite outer boundary. The non-semicircular shape of the outer boundary is chosen to facilitate the use of a very fine difference mesh upstream in the bow shock region and a coarse mesh downstream.

DESCRIPTION OF THE COMPUTATIONS

Five steady-state numerical solutions are presented for M_∞ = 2.0 flow past circular cylinders at freestream Reynolds numbers of 15.0, 46.8 and 157.2, based on cylinder diameter. The Reynolds number of the flow was varied by specifying different cylinder sizes and freestream flow conditions. We reference each solution by its approximate freestream Reynolds number with the three cases for Re_d = 46.8 differentiated by a suffix (i.e. 15, 47-1, 47-2, 47-3, 157).

The difference mesh used for our computations was constructed so that the angular spacing remained constant at $\Delta\theta$ = 6° for all cases. The length of a mesh element in the radial direction varied with the different computations. The smallest mesh was employed in case 47-1 for flow past a 3.056×10^{-3} ft diameter cylinder with $\Delta R = 1.030 \times 10^{-4}$ ft along the stagnation streamline. The largest mesh occurred in calculations such as case 47-3 which used the mapping (10). In such cases the mesh elements farthest from the cylinder stretched from some finite radial coordinate to infinity. The radial mesh variation was chosen by specifying α in transformation (10) so that there would be several mesh points per freestream mean free path across the bow shock at all times.

Three steady-state numerical solutions were obtained for Re_d = 46.8. The first case, 47-1, was started at time zero and run until the steady-state solution was reached at τ = 26.2 after approximately 3100 time steps. The non-dimensional characteristic time τ represents the number of cylinder diameters traveled by a particle moving with freestream velocity in a given time ($\tau = tV_\infty/2R_o$). The second case, 47-2, was computed to evaluate the effect of specifying freestream conditions along the finite outer boundary. Case 47-2 was initialized at τ = 26.2 by interpolating from the steady-state flow field of case 47-1. The governing equations were integrated in time until a steady solution was achieved at τ = 42.1 after approximately 2200 additional time steps. The third case, 47-3, was started at time zero and run to the steady state at τ = 39.1. This case required approximately 2900 time steps. Case 47-3 employed the coordinate mapping (10).

We choose the steady state from observations of the time-dependent flow in the bow shock and wake regions. At time zero the bow shock forms adjacent to the cylinder and then propagates upstream, gradually reaching a standoff distance which does not change appreciably with time. The wake flow then establishes itself after the bow shock has reached its steady-state position. After the bow shock is stationary we monitor the wake flow until this flow is invariant, or almost so, after several hundred time steps. The flow field at this stage is then said to represent the steady-state solution.

Figure 3 shows a comparison of the three numerical solutions for Re_d = 46.8 in the wake. Near the cylinder (R/R_o < 3) we see that the first two cases are in agreement. Case 47-3 predicts higher densities near the cylinder than the other cases due to the symmetry condition used downstream for the density. Far downstream of the cylinder, near the outer boundary, the first two cases predict large velocity, density and temperature gradients which are typical of those in rarefaction waves. These gradients result from the acceleration given the flow in the first two cases by the freestream velocity boundary conditions specified along the finite outer boundary. Case 47-3 predicts a more realistic qualitative behavior far downstream.

Figure 4 shows the density distribution across the bow shock for Re_d = 46.8. In this figure λ_o is the freestream gas mean free path. All three numerical solutions predict shock thicknesses of approximately three mean free paths. The analytical solution of Morduchow and Libby [9] for a viscous normal shock, shown in Figure 4, has a thickness of two mean free paths.

Figure 5 shows the time history of the density ratio across the bow shock at its intersection with the stagnation streamline. The quantity ρ_2 in this figure is the density behind the bow shock. Case 47-1 predicts a steady-state density ratio which is seven percent higher than that predicted analytically. The analytical solution of Chow and Ting [10] for a normal shock with first-order corrections accounting for shock curvature is used for comparison. Case 47-3 predicts a steady-state density ratio which is 14 percent higher than this same analytical solution.

None of the calculations for Re_d = 46.8 predicted steady-state separation as we might expect on the basis of incompressible experiments by Taneda [11]. Taneda showed that symmetric separation is unstable for Reynolds numbers above approximately 45 in incompressible flow. A calculation for Re_d = 15.0 was run to determine if the downstream symmetry conditions were compatible with the stability of the flow field. Case 15 was started at time zero and run until the steady state was reached at τ = 47.7 after approximately 3000 time steps. Case 15 showed no tendency toward separation and it thus appears that the symmetry boundary conditions do not contradict stability considerations at these low Reynolds numbers.

The density ratio across the bow shock predicted by case 15 was only three percent higher than the analytical solution of Chow and Ting [10]. Case 15 had a greater density of mesh points across the bow shock than any of the calculations for Re_d = 46.8. This case used almost four mesh points per freestream mean free path in the vicinity of the bow shock. This case shows that accurate prediction of the bow shock density ratio can be obtained if a fine mesh is used across the bow shock.

A numerical calculation was initialized for Re_d = 157.2 by interpolating values from the steady-state flow field reached in case 47-3. Approximately 1500 time steps were needed to achieve a steady-state solution. The numerical solution for Re_d = 157.2 predicted a bow shock standoff distance of Δ/R_o = 1.282. Figure 6 shows a comparison of the time history of the standoff distance for each calculation. The early time history of a calculation for Re_d = 704.6 is also shown in this figure. Case 705 was not completed because of the large amount of computer time required. Our numerical solutions predict that the standoff distance increases with decreasing Reynolds number as can be seen from this same figure.

Figure 7 shows the pressure coefficient predicted throughout the flow field in the numerical solution for Re_d = 157.2. Upstream of the cylinder the bow shock is quite pronounced. By the time the flow reaches X/R_o = 1 the bow shock has been smeared over many mesh widths. The dashed curves represent the locus of points of maximum positive pressure gradient. Pressure increases as we cross the dashed curves moving downstream with the flow. The numerical solution for Re_d = 157.2 predicts "shock-like" gradients in the wake which may represent the wake shock in the physical flow field. The calculations for Re_d = 15.0 and 46.8 did not predict a wake shock.

The numerical solution for case 157 predicted a small recirculation region downstream of the cylinder. Separation occurs at $\theta \approx 158°$ in our numerical solution for Re_d = 157.2. This separation is very much delayed from that observed in incompressible flow [11]. The numerical computations for Re_d = 15.0 and 46.8 did not predict separated flow.

EXPERIMENTAL INVESTIGATION

An experimental investigation of supersonic flow past cylinders was conducted in Supersonic Wind Tunnel No. 1 at the Ballistic Research Laboratories. The tests were conducted at a nominal Mach number of 2.00 and a nominal Reynolds number of 0.94×10^6 per foot. The models used for these experiments were cylindrical tungsten wires with diameters of 0.001, 0.002 and 0.008 inch, and a steel rod of 0.203 inch diameter. The freestream Reynolds number for these model sizes is 79, 157, 625 and 15,900, respectively. All models were located along the horizontal centerplane of the test section and spanned the test section through holes drilled in the tunnel windows or window frames. Tension was maintained on the wire models by the use of weights which were hung on one end of the wire; the other end was held fixed outside the opposite side of the tunnel.

Measurements were taken at several stations downstream of the 0.002 and 0.008 inch diameter models using a constant-temperature hot-wire anemometer. The dc voltage and rms voltage fluctuation measurements, taken with the hot-wire anemometer, were supplemented by schlieren photographs and spark shadowgraphs of the flow field.

Figure 8 shows results which are typical of those obtained from our measurements with the hot-wire anemometer. The traces shown in this figure were taken approximately 35 diameters downstream of the 0.002 inch model at a freestream

Reynolds number of 157.2. The dc voltage measurements give an indication of the local mass flow rate. Local values of density and velocity cannot be obtained from this measurement alone, since the hot-wire calibration is dependent on both the local Mach number and Reynolds number.

The root mean square hot-wire trace of Figure 8 shows that there are six peaks in the rms voltage plot at this axial station. The quantity Y/D in Figure 8 is the number of cylinder diameters above the flow field centerline. The two outermost peaks represent the bow shock, the next two peaks represent the wake shock and the two innermost peaks locate the edges of the inner wake. Behrens [12] has used hot-wire fluctuation measurements similar to these to discuss wake instabilities. It is interesting to observe from this trace that the fluctuation amplitudes in the wake shock are much lower than those in the bow shock and the inner wake edges.

The photographs for Re_d = 157.2 were not detailed enough to reveal whether separation was present at this Reynolds number. On the basis of spark shadowgraphs of the flow field for the 0.001 inch diameter model at Re_d = 79 it appears probable that a wake shock is present at this Reynolds number also. No hot-wire measurements were taken for Re_d = 79 due to model breakage.

CONCLUSION

Steady-state numerical solutions of the full Navier-Stokes equations have been obtained for flow past a circular cylinder at Re_d = 15.0, 46.8 and 157.2 in an M_∞ = 2.0 freestream. These calculations show that the mapping technique is compatible with the numerical solution of initial-boundary-value problems formulated using the full Navier-Stokes equations. These computations demonstrate the feasibility of performing supersonic flow calculations in two dimensions with boundary conditions imposed at ± ∞. This mapping technique gives acceptable results downstream of the cylinder and eliminates the "rarefaction type" gradients which result if freestream conditions are specified at a finite distance downstream of the cylinder.

The numerical solutions for Re_d = 15.0 and 46.8 predict no wake shock in the flow field. Experimental measurements were not possible at these low Reynolds numbers. Experimental measurements at Re_d = 157.2 show that a wake shock is present for M_∞ = 2.0 flow past a cylinder. Our numerical computations for Re_d = 157.2 predict pressure and temperature gradients in the wake which appear to represent the wake shock present in the physical flow field. It is not clear, however, that the details of the wake region are accurately predicted by these calculations.

Our calculations show that the degree of error present in the bow shock computations can be reduced to acceptable levels if very fine difference meshes are used in those regions of the flow with large gradients. In our calculations three or four mesh points per freestream mean free path appear to be necessary in the bow shock region. The numerical results for Re_d = 157 show that it is possible to predict a bow shock, wake shock and separated flow region for supersonic flow past a cylinder based on the numerical integration of the full time-dependent Navier-Stokes equations.

ACKNOWLEDGMENT

The author expresses appreciation to C. C. Bush, P. Gordon and J. H. Spurk for many helpful suggestions concerning this study.

REFERENCES

1. Crocco, L., "Solving Numerically the Navier-Stokes Equations," Report 63SD1001, March 1964, General Electric Co., Philadelphia, Pa.

2. Thommen, H. U., "A Method for the Numerical Solution of the Complete Navier-Stokes Equations for Steady Flows," GDC-ERR-AN733, April 1965, General Dynamics, San Diego, Calif.

3. Scala, S. M. and Gordon, P., "Solution of the Time-Dependent Navier-Stokes Equations for the Flow Around a Circular Cylinder," AIAA Journal, Vol. 6, No. 5, May 1968, pp. 815-822.

4. Allen, J. S. and Cheng, S. I., "Numerical Solutions of the Compressible Navier-Stokes Equations for the Laminar Near Wake," The Physics of Fluids, Vol. 13, No. 1, January 1970, pp. 37-51.

5. Trulio, J. G. and Walitt, L., "Numerical Calculations of Viscous Compressible Flow Around a Stationary Cylinder," NASA CR-1465, January 1970.

6. Scala, S. M. and Gordon, P., "Solutions of the Navier-Stokes Equations for Viscous Supersonic Flows Adjacent to Isothermal and Adiabatic Surfaces," Report 69SD1001, April 1969, General Electric Co., Philadelphia, Pa.

7. Gordon, P., "Nonsymmetric Difference Equations," Journal of the Society for Industrial and Applied Mathematics, Vol. 13, No. 3, 1965, pp. 667-673.

8. Sills, J. A., "Transformations for Infinite Regions and Their Applications to Flow Problems," AIAA Journal, Vol. 7, No. 1, January 1969, pp. 117-123.

9. Morduchow, M. and Libby, P., "On a Complete Solution of the One-Dimensional Flow Equations of a Viscous, Heat Conducting Compressible Gas," J. Aero. Sci., November 1949, pp. 674-684.

10. Chow, R. R. and Ting, L., "Higher-Order Theory of Curved Shock," PIBAL Report No. 609, August 1960, Polytechnic Institute of Brooklyn, Brooklyn, N.Y.

11. Taneda, S., "Experimental Investigation of the Wakes Behind Cylinders and Plates at Low Reynolds Numbers," J. Phys. Soc. Japan, Vol. 11, No. 3, March 1956, pp. 302-307.

12. Behrens, W., "Far Wake Behind Cylinders at Hypersonic Speeds: II. Stability," AIAA Journal, Vol. 6, No. 2, February 1968, pp. 225-232.

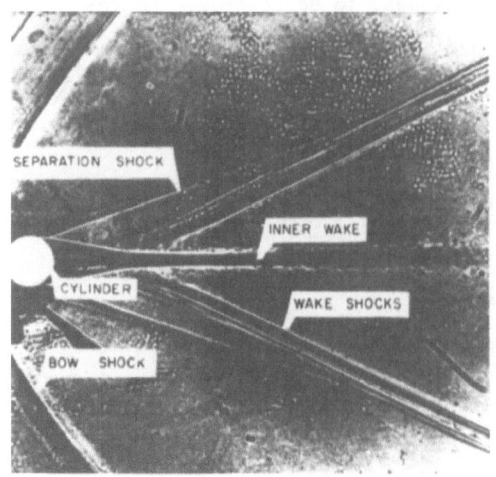

Fig. 1. Spark shadowgraph of flow field for $M_\infty = 2.0$ and $Re_d = 15,900$

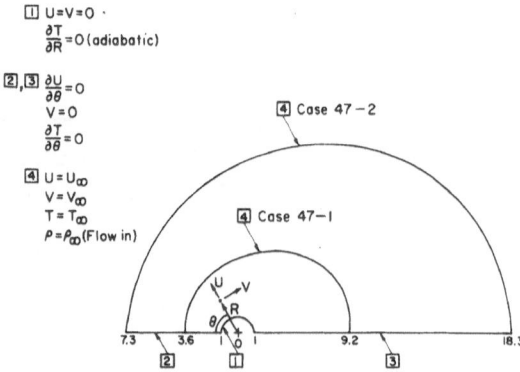

Fig. 2. Domain of computation and boundary conditions for two cases for $Re_d = 46.8$

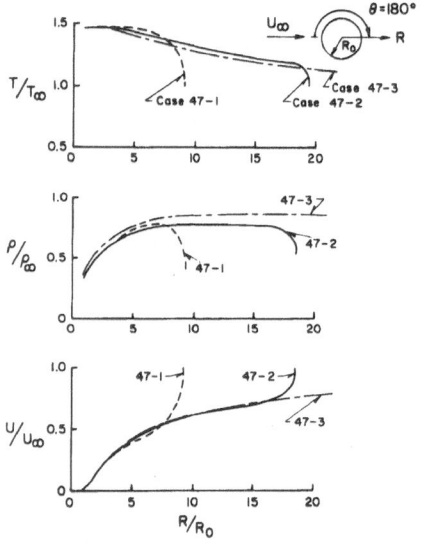

Fig. 3. Flow variables in wake ($\theta = 180°$) for $Re_d = 46.8$

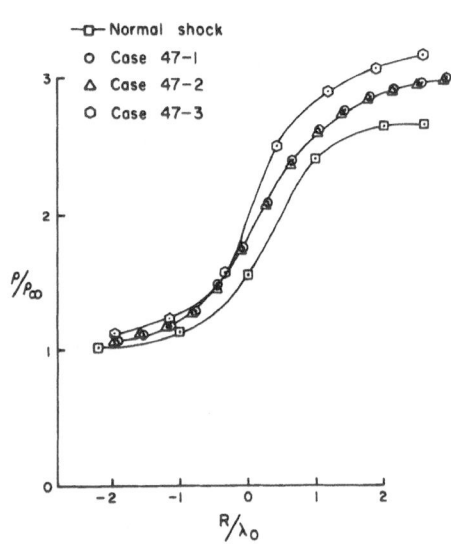

Fig. 4. Shock structure in terms of freestream mean free paths for $Re_d = 46.8$

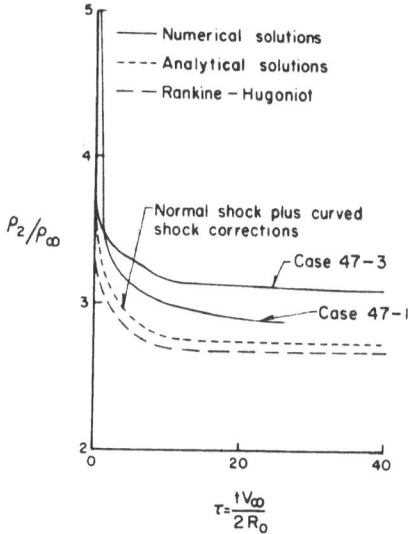

Fig. 5. Time histroy of density ratio across bow shock for $Re_d = 46.8$

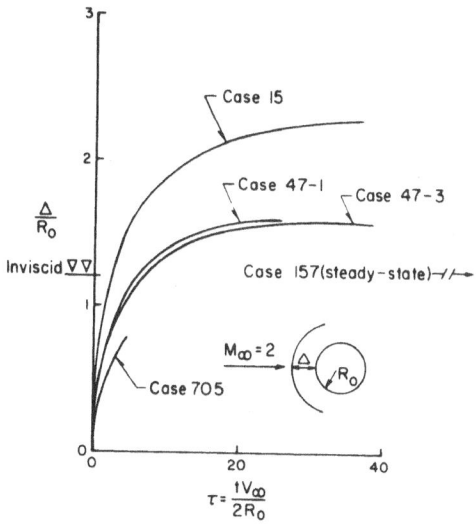

Fig. 6. Time history of bow shock standoff distance

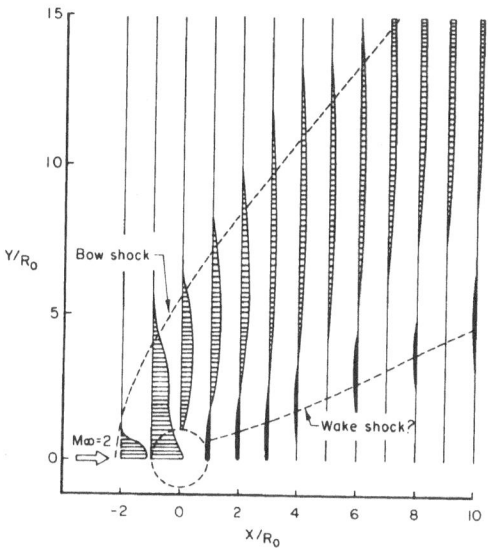

Fig. 7. Numerical solution for pressure coefficient for $Re_d = 157.2$

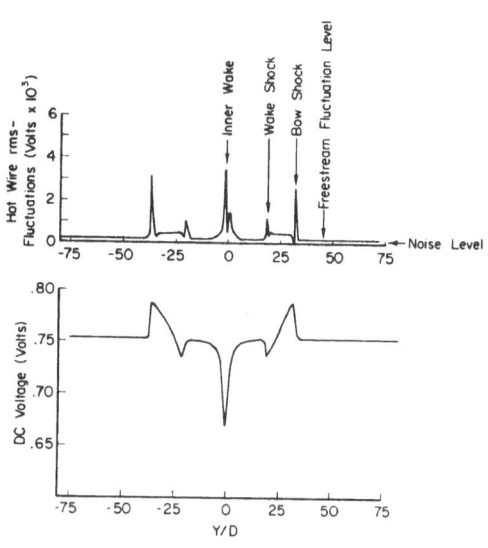

Fig. 8. Hot-wire survey at 35 diameters downstream for $Re_d = 157.2$

AN IMPROVED CONSTANT TIME TECHNIQUE FOR THE METHOD OF CHARACTERISTICS

C. A. Kot
IIT Research Institute, Chicago, Illinois

1. INTRODUCTION

The method of characteristics has been extensively applied for the solution of one-dimensional unsteady compressible flow problems. Since the method provides for the exact treatment of shocks and flow discontinuities, it is often the preferred approach when a detailed resolution of the flow field is required. Two different techniques are available when employing the method of characteristics. In both cases, the hyperbolic equations governing the flow are combined so that the resulting equations contain derivatives in one direction only. These directions define one-parameter families of curves called characteristics. One numerical technique (see Hoskin, 1970) relies on the solution at points formed by the intersection of opposite families of characteristics. To obtain information along constant time lines, a given spatial position, or along a particle path, requires therefore extensive two-dimensional interpolation. In the second, the so-called constant time approach, the characteristic equations are used to advance the solution in a mesh formed by lines of constant time and distance. In general this technique, first suggested by Hartree (1958), has been found to be more orderly and manageable for high speed computer application. In one particular version of this method (Chou et al., 1967), a mesh is formed of constant time lines and particle paths with the equations of motion expressed in the Eulerian variables. This method has been found to be flexible in its application and is the technique considered in the present paper. It is customary, with the constant time technique of the method of characteristics, to limit the time step by the Courant stability criterion (Courant et al, 1952). It will be shown that this restriction linking the time step to the mesh spacing is not mandatory. An approach in which spatial resolution and time step are chosen independently was developed and applied to the solution of blast wave fields. The latter technique was found to reduce the computational effort without a loss of accuracy. A difficulty often encountered with the method of characteristics is the slow convergence or even divergence of the iterative scheme at flow discontinuities, in particular at shock waves. A method for the advancement of such points, which relies on the appropriate combining of the finite difference equations and the partial algebraic solution of these equations is discussed in the present work. It was found to be always convergent, and significantly reduced the number of iterations required for a specified degree of convergence.

2. GOVERNING EQUATIONS

The equations of motion for one-dimensional unsteady flow in characteristic form may be written:

$$\frac{1}{\rho c}\,dp + du + \frac{\nu c u}{r}\cdot dt = 0 \qquad \text{on} \qquad dr = (u + c)\,dt \tag{1}$$

$$\frac{1}{\rho c}\,dp - du + \frac{\nu c u}{r}\,dt = 0 \qquad \text{on} \qquad dr = (u - c)\,dt \tag{2}$$

$$de + p\,dv = 0 \qquad \text{on} \qquad dr = u\,dt \tag{3}$$

where $r, t, p, \rho, v, e, u, c$ are respectively Eulerian coordinate, time, pressure, density, specific volume, specific internal energy, particle velocity and velocity of sound. The geometric factor $\nu = 0, 1, 2$ for plane cylindrical and spherical geometry.

For flow away from shock waves the set of equations is completed by a caloric equation of state of the form

$$F(p,\rho,e) = 0 \tag{4}$$

and the definition of the sound velocity. Shocks are treated as discontinuities, with the states across a shock related by the Rankine Hugoniot conservation equations.

The boundary conditions depend on the specific problem. Typically, they may consist of the condition of symmetry, material interfaces, a moving piston, etc. The computations are usually initialized by data obtained by auxiliary calculations such as the detonation of an explosive, an initial piston velocity, or a blast wave field generated from a similarity solution.

3. NUMERICAL PROCEDURE

In the constant time technique of the method of characteristics, a given number of grid points representing particles is advanced from one constant-time line to the next line. The time increment, Δt between the lines is predetermined and as mentioned earlier, is not limited by the local Courant stability criterion. Finite difference forms of the characteristic equations are used to compute the new point locations and their flow variables. It should be noted that in this approach, while the equations are written in Eulerian variables, one is essentially following the advancements of particles; thus a Lagrangian viewpoint is taken.

3.1 Calculation of a Typical Point

The finite-difference equations and the calculation procedure are best illustrated by the computations required for the advancement of a grid point or particle. Figure 1a shows grid point i which is to be advanced from the time t to the time $t + \Delta t$. It is assumed that all the grid points and the values of the variables at those points are known at time t. Point j indicates the location of the particle i at time $t + \Delta t$. Points a and b are the intersections of the u + c and u - c characteristics through point j with the constant-time line t.

The characteristic equations are written in finite-difference form using first order differences with centered values. Typical of these equations are the expressions applicable along the (u + c) characteristic for point j, which are given below.

$$r_j = r_a + (\overline{u} + \overline{c})_{aj} \, \Delta t \tag{5}$$

$$u_j = u_a - (p_j - p_a)\left(\frac{1}{\overline{\rho}\,\overline{c}}\right)_{aj} - \nu\left(\frac{\overline{c}\,\overline{u}}{\overline{r}}\right)_{aj} \Delta t \tag{6}$$

The quantities designated by a bar represent average values. Variables at points a and b on the constant-time line t are obtained by interpolation between the known grid points on this line. Either first or second order interpolation may be used. The finite difference equations after rearrangement, together with the interpolation equations and the equations relating the state variables yield a set of equations which must be solved iteratively to determine point j.

First the new spatial locations of point j and the intercepts a and b are computed. The interpolation equations are then used to obtain the variables at a and b. Then the finite difference forms of equations (1) and (2) are combined to yield the pressure p_j. Finally the velocity u_j is determined from equation (6). After p_j and u_j have been found, during a particular iteration, the finite difference form of the energy equation (3) is solved simultaneously with the equation of state (4) to obtain the internal energy e_j and the density ρ_j. Obviously the sound velocity c_j can then be obtained from the appropriate thermodynamic relationships.

To start the iteration procedure, initial values of the variables p, ρ, u, and e at points j, a, and b must be known. These are taken to be the values at points i, $i-1$, and $i+1$, respectively. The iteration procedure is stopped when convergence criteria for both velocity and pressure are satisfied. Typically, a convergence to within 0.00001 of the values in the previous iteration may be required.

For points which are close to a discontinuity, the characteristic line through the point may intersect the discontinuity. This is shown in Figure 1c for the point p which is to be advanced to position q. Obviously, conditions at point b' cannot be used to compute point q, since they are on the far side of the discontinuity. The calculation procedure is modified to find point b on the boundary or discontinuity. Linear interpolation between states s and z is used.

3.2 Shock Point Calculation

Boundary points such as free and rigid surfaces, material interfaces, and shock waves must be treated by special procedures. In particular, the calculation of shock waves, which are considered true discontinuities, may afford some difficulties. Various schemes have been developed by other investigators (Hoskin, 1964; Huang, et al, 1966) for the treatment of shock waves. In general these procedures, while adequate, require considerably greater numbers of iterations for convergence than the schemes for the ordinary points. A method for the calculation of shock points which ensures rapid convergence was therefore developed. The approach is similar to the one used for ordinary points in that it relies on the partial algebraic solution of the shock and finite difference equations. The method is illustrated for a rightward facing shock in Figure 1d. The shock advances from position s to s' in the time Δt. State 1 refers to the conditions ahead of the shock while state 2 is behind the shock.

First, the position of the shock is estimated using the finite difference form of the expression $dR/dt = U$. Here R is the shock radius and U the shock velocity. The solution on the low pressure side of the shock (state 1) is then calculated by a method similar to that for an ordinary point. Flow variables at point z are obtained by interpolation. If the shock is propagating into a uniform region, conditions in state 1 are known a priori. The location of the intercept of the characteristic line behind the shock with the constant time line at point a is determined and the variables at the point obtained by interpolation. The finite difference compatibility equation for the characteristic through state 2 is the same as equation (6), where subscript j must be replaced by 2. Combining this expression with the Rankine-Hugoniot momentum equation

$$(u_2 - u_1)^2 = (p_2 - p_1)\left(\frac{1}{\rho_1} - \frac{1}{\rho_2}\right) \tag{7}$$

and solving for the pressure p_2, one obtains

$$p_2 = \frac{1}{B}\left\{A + Bp_a + \frac{D}{2B} + \sqrt{D\left[\frac{D}{4B^2} + \frac{A}{B} + (p_a - p_1)\right]}\right\} \tag{8}$$

where $A = u_a - u_1 - \nu\left(\frac{\overline{c}\,\overline{u}}{\overline{r}}\right)_{a2}$; $\quad B = \left(\frac{1}{\overline{\rho}\,\overline{c}}\right)_{a2}$ $\quad$ and $\quad D = \left(\frac{1}{\rho_1} - \frac{1}{\rho_2}\right)$

The particle velocity u_2 is then obtained from the compatibility equation, while the density $\hat{\rho}_2$ and the energy e_2 are calculated by the simultaneous solution of the shock energy equation and the equation of state. The sound velocity c_2 and the shock velocity U_2 are then readily calculated and the procedure is repeated with a new estimate of the location of point s'.

The technique was found to converge rapidly, even when the equation of state was in tabular form (Kot, 1970). The reduction in computational effort is quite significant. For simple shocks, convergence to within 10^{-5} of the previous value was typically achieved in five iterations, while the usual procedures required about 30 iterations.

4. TIME STEP, STABILITY AND ACCURACY

The time increment in a numerical scheme is often subject to certain restrictions because of stability considerations. For schemes based on the method of characteristics, it has been shown (see e.g. Chou et al, 1967) that stability is assured when the domain of dependence of any point as given by the finite difference equation is not less than the exact domain of dependence of the differential equation. On the other hand the Courant stability criterion, for a fixed grid finite difference scheme, requires that:

$$\Delta t \leq \frac{\Delta t}{c + |u|} \qquad (9)$$

The extension of the Courant criterion to schemes using particle paths as moving grids and employing the method of characteristics yields the following approximate time step restriction (see Chen and Chou, 1971).

$$\Delta t \leq \frac{\Delta r}{\bar{c}} \qquad (10)$$

where $\bar{c}$ is the sound velocity average between adjacent grid points. The time increment obtained in this manner is illustrated by $\Delta t'$ in Figure 1b where point i is advanced to location j'. Obviously the true domain of dependence of point j' in this scheme is the distance $\overline{a'b'}$ while the domain of dependence of the numerical scheme is the larger distance from point i-1 to point i+1, thus stability is assured. Further examination of the above expressions shows that the primary objective of the restrictions imposed on the time step is that the intercepts of the characteristic lines through point j' with the constant time line line t be within the spatial intervals formed by the point of interest and its immediate neighbors. This has the advantage that no search for the intercept points a' and b' is required. This approach has been widely used by other investigators (Chou et al, 1967 and Hoskin, 1964). It suffers, however, from a major shortcoming, namely, that a refinement or coarsening of the spatial distribution of points is accompanied by a corresponding change of the time step. The total computational effort with this method for a given problem varies therefore approximately as the square of the required spatial resolution.

To avoid this difficulty, the computational scheme outlined earlier (Section 3) was developed. In principle, this method permits the use of time steps of arbitrary size. Examining the stability of the scheme, we see (Figure 1a) that the true domain of dependence for point j is the distance $\overline{ab}$, while the domain of dependence of the finite difference method is the larger distance $\overline{mn}$. Thus the basic stability criterion is satisfied. The slight inconvenience of having to search for the intersection of the characteristic lines with the constant time line is far outweighed by the computational flexibility achieved through the uncoupling of the time step from the spatial resolution. That this is possible is not too surprising considering that the numerical method of characteristics is basically an implicit scheme.

The major restrictions imposed on the time step are those of accuracy. Using series expansion, it can be shown (Chou, et al, 1967) that the truncation errors which are introduced by replacing the differential equations by the finite-difference equations are of the order h^2 where h is some parameter proportional to arc length along the characteristic lines between points at which the solution is determined in the r,t-plane. Obviously, as the time step increases, so will the arc length along the characteristics and, in consequence, the truncation error.

Since the characteristic equations are nonlinear, no rigorous procedure for estimating the accuracy of the numerical method exists. The accuracy of the method developed was thus estimated by comparing its computations with the results of an exact solution. The similarity solution of a point source explosion in air obtained by Sedov (1959) was chosen for comparison. Numerical results of Sedov's solution were obtained to serve as a standard of comparison. To avoid the difficulties associated with the origin, the region surrounding that point was excluded. The calculated region was bounded by a constant-time (nonzero) line on the bottom, a back boundary on the left with the equation $r_b = 0.8R$, and the shock front on the right (see Figure 2). Here r_b is the radius of the back boundary and R is the shock radius. On the initial constant-time line, all variables are prescribed and given by the exact solution. On the back boundary at r_b, the particle velocity is specified for all times, again, as obtained from Sedov's exact solution. At the shock front, the strong-shock conditions are assumed to hold. The computations were performed with dimensionless variables. The ambient density, the initial shock pressure, and an explosion energy of unity were used to construct dimensionless variables. Figure 3 is a comparison of the accuracies of the shock pressure, particle velocities, and shock radii for four computations. Two of these were carried out employing time steps consistent with equation (10), i.e., using the usually accepted computational procedure. Of these, Case 1 used 21 spatial intervals on the starting line and Case 2 used 41 spatial intervals. In Cases 3 and 4, computations were started with 41 spatial intervals and the time steps were equal to $2\Delta t'$ and $4\Delta t'$ respectively, where $\Delta t'$ is the time step dictated by equation (10). It may be observed that, for all cases, the accuracies are quite good. However, the method developed here proves its superiority since it provides comparable accuracy with less computation time. Thus, Case 3 requires only one-half the time of Case 2, and Case 4 provides, again, a time reduction by a factor of two. In fact, Case 4 requires as little time as Case 1 and is more accurate at the shock front. Figure 4 shows the error profiles for the entire computed region for the four cases. Again, the accuracies obtained in Cases 3 and 4 are quite good and comparable to those obtained by the conventional technique. At the time of this comparison, the shock pressure has decayed to four-tenths of its starting value. All the errors are determined by comparison to Sedov's exact solution.

The accuracy of the computational procedure at late times (many time steps) was checked in a calculation with a time step between $2\Delta t'$ and $3\Delta t'$. The computation was terminated when the shock pressure decayed to approximately 0.5 percent of its original value. This required 290 time steps starting again with 41 spatial intervals. At the end of the computation, the number of spatial intervals had increased to around 120, because of an increase in the shock-engulfed region. The percent error in the shock pressure, particle velocity, and shock radius is given in Figure 5. Again, the errors are quite small and approach steady values for late times. The error in the pressure, velocity, sound-speed profiles at early times (10 time steps) is given in Figure 6, while for the final time, the error profiles are shown in Figure 7. It is gratifying to note that as time increases and the slopes of the variable profiles in the blast wave become less steep, the computational error decreases. Thus, the computation method employed is quite stable and accurate and permits considerable reduction of the computation effort at the same time. Application of the computational scheme to more complex problems, in particular in the treatment of underwater explosion phenomena has also been successful (Kot, 1970).

ACKNOWLEDGMENT: This work was partially supported by ONR (Contract N00228-67-C2717), AFOSR (Contract F44620-71-C-0060) and IIT Research Institute.

REFERENCES

Chen, S. and Chou, P. C., "A Method of Characteristic Code for Energy Deposition Calculations," DIT Report No. 125-15, (March 1971).

Chou, P. C., Karpp, R. R. and Huang, S. L., AIAA Journal, 5, 618-623, (1967).

Courant, R., Isaacson, E., and Rees, M., Comm. Pure and Applied Math., 5, 243-255, (1952).

Hartree, D. R., Numerical Analysis, Oxford University Press, (1958).

Hoskin, N. E., "Solution by Characteristics of the Equations of One-Dimensional Unsteady Flow," Methods of Computational Physics, 265-293, Academic Press, New York, (1964).

Hoskin, N. E. and Lambourn, B. D., "The Computation of General Problems in One Dimensional Unsteady Flow by the Method of Characteristics," Proceedings of the Second International Conference on Numerical Methods in Fluid Dynamics, 230-235, (September 15-19, 1970).

Huang, S. L., and Chou, P. C., "Solution of Blast Waves by a Constant Time Scheme in the Method of Characteristics," DIT Report No. 125-9, (August 1966).

Kot, C. A., "Point Source Underwater Explosions," Ph.D. Thesis, Illinois Institute of Technology, (1970).

Sedov, L. I., Similarity and Dimensional Methods in Mechanics, Academic Press, New York, (1959).

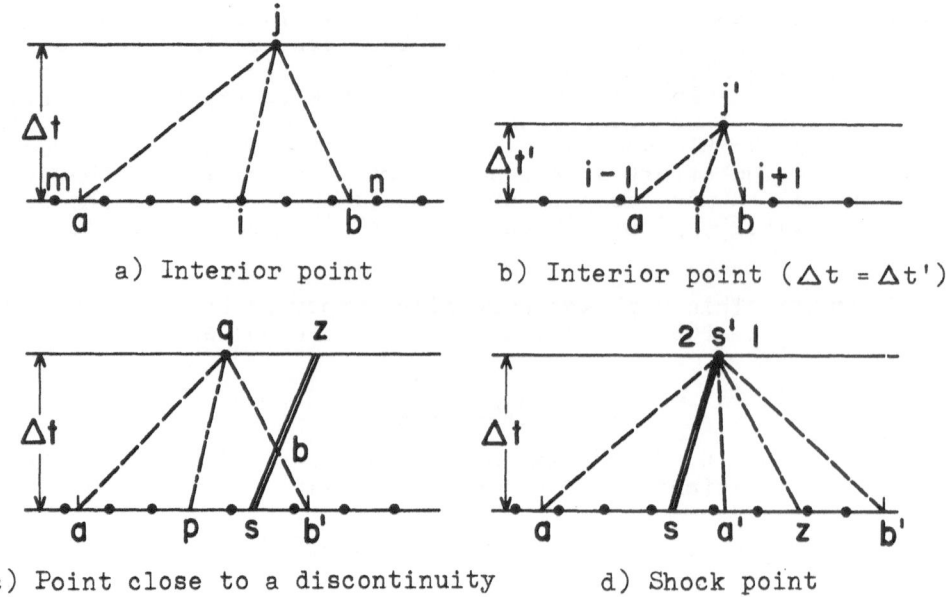

a) Interior point   b) Interior point ($\Delta t = \Delta t'$)

c) Point close to a discontinuity   d) Shock point

Fig. 1. Grid constructions for typical points

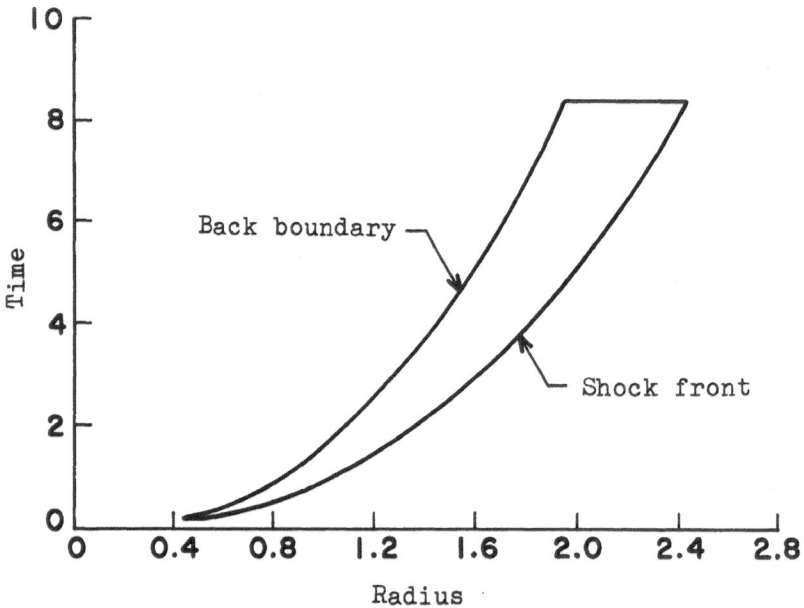

Fig. 2. Spherical blast wave region used in accuracy test

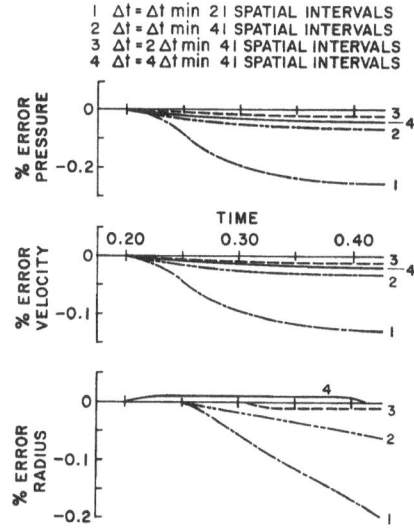

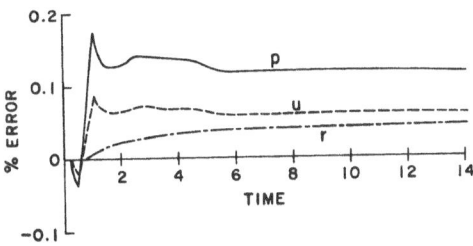

Fig. 5. Error along shock front
($\Delta t = 2.5 \, \Delta t$ min)

Fig. 3. Error comparison at
shock front

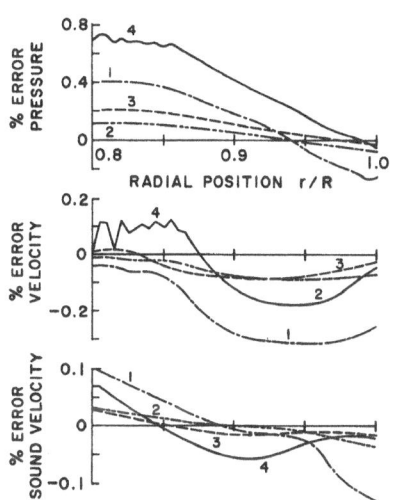

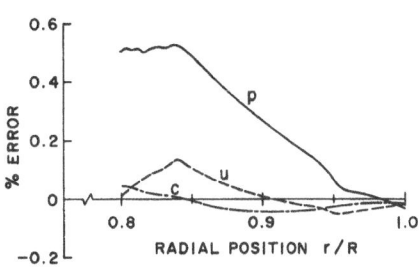

Fig. 6. Error profiles after 10
time steps

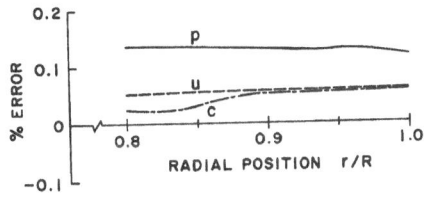

Fig. 4. Comparison of error
profiles

Fig. 7. Error profiles after 290
time steps

FINITE DIFFERENCE METHODS FOR THE
STEADY-STATE NAVIER-STOKES EQUATIONS

Patrick J. Roache, Theoretical Fluid Dynamics Division
Sandia Laboratories, Albuquerque, New Mexico, USA

ABSTRACT

Two iterative methods for numerically solving the incompressible 2D steady-state Navier-Stokes equation are presented. These are the Numerical Oseen (NOS) method and the Laplacian Driver (LAD) method. Unlike most methods, these are not time-dependent or even time-like in their iterations. The methods make use of recent advances in numerically solving 2D linear second-order partial differential equations with methods which are direct (i.e., non-iterative).

INTRODUCTION

Most numerical methods for solving the steady-state Navier-Stokes equations of fluid dynamics are time-dependent methods, in which the steady-state solution (if it exists) is obtained as an asymptotic time limit. These methods have the advantage of not requiring the assumption, often unjustified, that a steady-state solution exists. Even methods which are not consciously designed to model the time-dependent equations are usually time-like in their iterations. An exception is the excellent method of Davis (1972), in which each complete iteration cycle is split into a steady-state boundary layer solution followed by a linear correction. Although the assumption of a steady-state condition is required, a great advantage of computational speed accrues. Indeed, at high Reynolds number R_e, the solution for flow over a parabola is virtually attained in one iteration, via the boundary layer solution.

We would like to devise steady-state methods which are faster than time-like methods, but which are applicable to problems in which boundary layer solutions are not even qualitatively correct. In the present work, two new iterative methods for solving the incompressible 2D steady-state Navier-Stokes equations in stream function (ψ)-vorticity (ζ) form are presented and tested. These are not time-dependent or even time-like in their iterations. They make use of recent advances in solving 2D linear second order partial differential equations using methods which are direct (i.e., non-iterative).

The governing equations which we consider are the vorticity transport equation, the Poisson equation for stream function, and the relations between ψ and the velocity components u and v. Subscripts indicate partial derivatives, as in $\zeta_t = \partial\zeta/\partial t$.

$$\zeta_t = -R_e \, \nabla\cdot\vec{V}\zeta + \nabla^2\zeta \qquad (1a)$$

$$\nabla^2\psi = \zeta \qquad (1b)$$

$$\vec{V} = u\vec{i} + v\vec{j} = \psi_y\vec{i} - \psi_x\vec{j} \qquad (1c)$$

R_e is the Reynolds number. The interior equations (1) are supplemented by boundary conditions proper to a particular problem.

TWO METHODS

The first method is a Picard-type method. It is similar to an n-th order Oseen approximation (Schlichting, 1968) solved numerically. We refer to it as the Numerical Oseen method and, following custom, we abbreviate it with a 3-letter acronym NOS.

After an initial guess on ψ at all points, the iterative cycle begins with the boundary evaluation of ζ. Then the steady-state form of (1a) is solved for the k-th iteration, using $O(\Delta x^2)$ centered differences, as

$$0 = - R_e \ \nabla \cdot (\vec{V}^{k-1} \zeta^k) + \nabla^2 \zeta^k \tag{2}$$

The direct solution of (2) is obtained using the EVP method (Roache, 1971) or possibly another general solver for general second-order linear finite difference equations. Then the Poisson equation (1b) is solved for ψ (by a direct method) and the cycle is repeated.

It is worth emphasizing that this method is not time-like. At each iteration, the linearized vorticity transport equation is solved exactly for the steady-state solution.

This numerical Oseen method requires the solution of a second-order linear FDE with first-order terms. Since the coefficients of these terms (u and v) vary with each iteration, the EVP solution must be initiated each time, i.e., a new influence coefficient matrix must be generated and inverted. This is in contrast to the Poisson solution, which need be initiated only once. Also, the applicability is limited by the mesh restrictions of the EVP method (Roache 1971).

A second method, which avoids these shortcomings, is obtained by replacing the vorticity iteration of NOS by

$$\nabla^2 \zeta^k = F^{k-1} \equiv R_e \ \nabla \cdot (\vec{V}^{k-1} \zeta^{k-1}) \tag{3}$$

In this method, only the Laplacian term of the vorticity transport equation drives the iterations toward a solution; the method is accordingly described as the Laplacian Driver method and abbreviated as LAD. Other aspects are the same as NOS. This method may use EVP or other direct Poisson solvers such as Fourier methods or Buneman's method (Dorr, 1970; Buneman, 1969).

DRIVEN CAVITY PROBLEM

These methods were tested and compared to a time-dependent method for the driven cavity problem, as shown in Figure 1. This is a geometrically simple problem, and has often been used as a test for numerical methods. The continuum boundary conditions for this problem are simple and unambiguous, compared to flow-through problems. On all boundaries, the no-slip conditions prevail, giving $\psi = 0$ and allowing the evaluation of ζ by any of three equations. We write the equations for the moving "lid" at $j = J$. For derivations and references, see Roache (1972b).

$$\zeta_{i,J} = \frac{2(\psi_{i,J-1} - \psi_{i,J} + \Delta y \cdot U)}{\Delta y^2} + O(\Delta y) \tag{4a}$$

$$\zeta_{i,J} = \frac{3(\psi_{i,J-1} - \psi_{i,J} + \Delta y \cdot U)}{\Delta y^2} - \frac{1}{2} \zeta_{i,J-1} + O(\Delta y^2) \tag{4b}$$

$$\zeta_{i,J} = \frac{-\psi_{i,J-2} + 8\psi_{i,J-1} - 7\psi_{i,J} + 6\Delta y \cdot U}{2\Delta y^2} + O(\Delta y^2) \tag{4c}$$

The first-order method (4a) is most commonly used. The method (4b) is due to Woods. Jensen's method (4c) is obtained by passing a cubic equation for ψ through the wall point and two neighboring points. Briley (1970) has shown in time-dependent calculations using (4c) that the stability and accuracy of the total solution is enhanced if the velocities near the wall are also evaluated in a form compatible with the assumed cubic form of ψ. Thus, we evaluate u near the lid as

$$u_{i,J-1} = \frac{-\psi_{i,J-2} - 4\psi_{i,J-1} + 5\psi_{i,J}}{4\Delta y} - \frac{1}{2} u_{i,J} \tag{5}$$

Equations similar to (4) and (5) apply at the other walls. Since R_e is based on the lid velocity, we set $u_{i,J} \equiv U = 1$.

Both methods were unstable at all R_e tested, including $R_e = 0$. It was necessary to under-relax the boundary evaluations of vorticity (ζ_b) according to

$$\zeta_b^{k+1} = f \cdot r + \zeta_b^k (1 - r) \qquad (6)$$

where f is a function indicated in (4) and r is the under-relaxation parameter, $0 < r < 1$. It was not necessary to under-relax interior values of ζ or ψ.

RESULTS OF THE DRIVEN CAVITY CALCULATION

Numerical tests were run on a small problem (10 x 10 cells, or $I = J = 11$ and $\Delta x = \Delta y = 1/10$). For comparison, the simplest time-dependent method was also tested; it uses forward-time and centered-space (FTCS) differences, and is $O(\Delta t, \Delta x^2)$. The complete tests varied the method (NOS, LAD, and FTCS), the wall vorticity equation (4a, b and c), the under-relaxation parameter r ($0 < r \leq 1$) and the flow parameter R_e ($0 \leq R_e \leq 100$). All cases were carried out to 111 iterations, unless strong instability occurred. Print-outs of all field variables and several diagnostic functionals were examined subjectively for iteration convergence. In some cases (first-order wall ζ, low R_e and near-optimum r) iteration convergence was unequivocal, with floating-point zeroes for the change in quantities indicating convergence to the single-precision accuracy of the machine (a CDC 6600 with ≈ 14 decimal significant figures).

In order to quantify the comparisons, we have chosen to consider $\sum_{ij} |\zeta_{ij}|$ where the summation extends only over the internal points $2 \leq i \leq I-1$, $2 \leq j \leq J-1$. Then

$$DQ^{k+1} = \sum_{ij} |\zeta_{ij}^{k+1}| - \sum_{ij} |\zeta_{ij}^k| \qquad (7)$$

was tested. We plotted the number of iterations K required to obtain $|DQ^K|/r < 10^{-4}$. For the present problems, this is approximately equivalent to requiring a normalized value of $|DQ^K| / \left(r \cdot |\sum_{ij} \zeta_{ij}^K| \right) \approx 10^{-6}$.

The results for $R_e = 20$ using the first-order wall ζ equation (4a) are shown in Figure 2. The abscissa for the FTCS method is $\Delta t / \Delta t_{crit}$. At this R_e, the Δt_{crit} is controlled by a diffusion limitation rather than by a local Courant number limitation.

This Figure 2 shows that for the driven cavity problem, the NOS and LAD methods are not significantly different in their convergence properties, and that both exhibit an expected optimum value of r, which we denote by r_0. Beyond the vaguely-defined r_0, the convergence rate deteriorates rapidly (more so for LAD than NOS) and the methods then become unstable. This behavior is similar to the FTCS dependence on $\Delta t / \Delta t_{crit}$.

The results for other parameters varied are not sufficiently distinct to warrant a full presentation of data; the following verbal description will suffice. At $R_e = 0$, $r_0 \simeq \frac{1}{4}$ for the first-order wall ζ evaluation (4a), and about 1/5 for the second-order equations (4b,c). As R_e increases, r_0 increases slightly for NOS. For $R_e = 100$, convergence is not obtained for any of the methods, including FTCS. The use of the second-order methods (4b,c) for wall ζ caused faster instability for $r > r_0$, and generally slowed convergence slightly for all r.

It is worth emphasizing that when iteration convergence did occur, all three methods (NOS, LAD, and FTCS) did converge to the same answer, independent of r and Δt.

RELATIVE IMPORTANCE OF LAGGING BOUNDARY CONDITIONS

From the results of Figure 2, it is clear that both new methods are superior to the time-dependent FTCS method, for the convergence criterion used. However, the improvement is not at all as large as expected. (Note that the use of a second-order time-dependent method may have speeded the time-dependent convergence.) It was also surprising that the NOS method, which includes advance information in the advection term, was not clearly faster (in the number of iterations) than LAD. Both of these observations suggested the experiment described below.

There are two sources of coupling between the ψ and ζ interior equations. The first and most obvious source is the lagging velocity field $\lceil u, v = fcn (\psi) \rceil$ in the advection term of the NOS method, and the velocity field and ζ field in the lagging advection term F of the LAD method. The second source is the boundary condition (4) on ζ, through which ζ_b depends on internal values of ψ. Because of the small difference between the behavior of NOS and LAD (which treat the nonlinear advection term differently) and between cases with $R_e > 0$ compared to $R_e = 0$ (which has no advection term) it was clear that the lagging boundary values of ζ were a major contributor to the slower-than-expected convergence.

To qualitatively determine the importance of the lagging boundary conditions, the following procedure was followed. A solution for a particular R_e was determined, starting with $\psi = 0$ and $\zeta = 0$ everywhere, using $r \simeq r_0$ and a stringent convergence criterion such as $DQ/r < 10^{-10}$. Then the problem was run again, starting with $\psi = 0$ and $\zeta = 0$ at all internal points, but with the boundary values of ζ frozen at their correct steady-state values obtained in the first solution. For $R_e = 0$, the final results are obtained immediately. For $R_e = 20$, the results are plotted in Figure 2 as a horizontal line. (Since boundary values of ζ are fixed and since the under-relaxation factor r is applied only to the boundary ζ, the results do not depend on r.) The reason for defining $\sum_{ij} |\zeta_{ij}|$ only over internal points is now clear; if the summation had included boundary points, the convergence criterion would have been biased in favor of these experiments for which the boundary values are frozen.

The results in Figure 2 show very clearly that the lagging boundary values of ζ, rather than the nonlinear advection term, are responsible for the lower than expected performance of the methods. With frozen boundary values of ζ, convergence was obtained at K = 12 for LAD and K = 14 for NOS. Using (4c) for wall ζ with LAD, convergence occurred at K = 10. (By contrast, the convergence rate of the FTCS method actually deteriorates when frozen boundary values are used. This serves to emphasize again the distinction between the present methods and time-like methods.)

DIFFERENT BOUNDARY EVALUATIONS

The effect of using second-order evaluations of ζ is to increase formal accuracy but to deteriorate stability, especially for $r > r_0$. In Figure 3, we have plotted four evaluations of ζ along the driving lid; one from each of equations (4a, b, c) in the coarse mesh (10 x 10 cells) and one from a reference solution in a fine mesh (128 x 128 cells). The significance of the results is clouded because of the singularities in ζ at the lid corners. We can generally recommend the second-order methods over the first-order method for this problem. (For some 1D problems, the first-order method is clearly superior. See Roache, 1972a.) There is little difference between the two second-order methods. Since Woods' method (4b) is already consistent with the usual centered-difference evaluation of velocities near the wall, it is recommended over Jensen's method (4c).

LARGE PROBLEMS

An advantage of both the NOS and LAD methods is that the number of iterations K required to reach convergence is approximately independent of problem size (for small Δx). However, the optimum relaxation factor r_0 is mesh dependent. Analysis of simple 1D problems (Roache 1972a) shows that r_o varies inversely with Δx. Estimates of r_o may be obtained from a series of coarse-mesh solutions, or from fine-mesh solutions at different R_e.

The reference solution ($R_e = 20$) of Figure 3 was determined with LAD using a version of Buneman's (1969) Poisson solver and Woods' method (4b) for the wall vorticity. This large problem (128 x 128 cells) was solved to convergence (1 part in 10^4 for ζ in the center of the lid) in 38 complete iterations, each of which is comparable in machine time to a complete time-step in an explicit time-like method. The solution mesh was fine enough to disclose two weak counter-rotating vortexes in the lower corners. The entire solution, including some print-out, required about 4 minutes on a CDC 6600.

DRIVEN CAVITY WITH SYMMETRIES

The symmetric half of a double driven cavity is identical to the single driven cavity problem of Figure 1, except that $\zeta = 0$ along the side of the mesh ($J = 1$) opposite the lid. Although no iteration is necessary for ζ along this one boundary $J = 1$, the improvement in the convergence rate is insignificant. The iterations become unstable at a somewhat lower R_e than for the single driven cavity, since the lack of one no-slip boundary increased the speed of circulation at the internal points.

More contrived problems allow for 2 or 3 symmetry boundaries with $\zeta = 0$. The triple-symmetry problem does indeed converge much faster. At $R_e = 20$ with $\Delta x = 1/10$, convergence to $DQ/r < 10^{-4}$ was achieved at as low as K = 15 with NOS with first-order wall ζ.

FLOW-THROUGH PROBLEM

A driven cavity problem is diffusion dominated. The R_e is based on the lid speed, but the speed at interior points is much less. It appeared that the NOS method, which includes advance information in the advection term, might be significantly better than LAD for a flow-through problem wherein the internal speeds are larger.

A developing channel flow (Figure 4) was chosen for comparison. The inflow boundaries ψ and ζ were fixed with a linear u profile and v = 0 (Couette flow). The maximum value u = 2 was chosen so that R_e is based on the bulk mixing velocity. Thus, $\psi(y) = y^2$ and $\zeta(y) = 2$ at inflow. At y = 1, the upper boundary of the symmetric lower half of the problem, we have $\psi = 1$ and $\zeta = 0$. Along the lower boundary, no-slip wall conditions apply, with $\psi = 0$ and ζ given by (4).

The outflow boundary at i = I was set by methods used in time-dependent methods (Roache, 1972b). Convergence was obtained using

$$\partial^2\psi/\partial x^2 = 0 \qquad (8a)$$

$$\partial\zeta/\partial x = 0 \qquad (8b)$$

The condition (8a) was set at i = I, allowing the solution of an ordinary finite difference equation for ψ (Roache, 1970). We used $\Delta y = 1/10$ and a mesh aspect ratio $\Delta x/\Delta y = 10$. Convergence results for $R_e = 1$ and $R_e = 50$, using Woods' equation (4b) for wall ζ are presented in Figure 5. As expected, the presence of three boundaries on which the steady-state values of ψ and ζ are known did significantly speed convergence, compared to the driven cavity problem. Also, NOS did perform significantly better than LAD in the number of iterations required.

High R_e solutions are not possible with fixed outflow ζ, because of the cell Reynolds number limitation. This limitation of $uR_e\Delta x < 2$ is common to all finite difference methods which use centered differences (Roache, 1972b,c). But with the gradient conditions (8) used at outflow, the solution obtained at $\max(u\text{Re}\Delta x) \approx 100$ was smooth.

Optimum values of r_0 were predicted (Roache, 1972a) from the analysis of two 1D problems, Couette flow and planar Poiseuille flow. The r_0 for Poiseuille flow, r_{op}, is shown in Figure 5. For low Re, the asymptotic Poiseuille solution develops quickly; r_{op} is appropriate, and accurately predicts r_0 for the NOS method at Re = 1. For high Re, the inflow Couette flow profile persists. However, the r_0 for Couette flow does not accurately indicate r_0 for Re = 50 because the centerline value of vorticity for this 2D problem is $\zeta_{i,J} = 0$, rather than the Couette-flow value of $\zeta_{i,J} = 2$.

CHOICE BETWEEN NOS AND LAD

For the driven cavity problems, the important difference between the NOS and LAD methods is not so much in the number of iterations K required to meet a convergence requirement, but in the computation time per iteration. Since LAD involves only the solution of a Poisson equation, it is faster than NOS (which requires the generation and inversion of a new influence coefficient matrix at each iteration) beyond the first iteration. This difference is especially important in large problems. Also, the direct Poisson solvers required by LAD are more generally available than the EVP method required by NOS.

For the flow-through problem of developing channel flow, NOS was significantly faster in number of iterations than LAD. For relatively small problems such as the 10 x 10 cell problem tested, NOS is also faster in computation time. But for large enough problems, the faster computation speed of LAD per iteration will offset the greater number of iterations, and LAD may still be preferred over NOS for large flow-through problems.

A SUGGESTED THIRD METHOD

The results of the numerical experiments cited show the importance of the lagging boundary values of ζ. It appears that this difficulty could be overcome by the use of a direct solver for the inhomogeneous biharmonic equation.

Substituting (1b) into (1a) gives, with $\nabla^2\nabla^2\psi \equiv \nabla^4\psi$,

$$\frac{\partial}{\partial t}(\nabla^2\psi) = -R_e \nabla\cdot(\vec{V}\nabla^2\psi) + \nabla^4\psi \tag{9}$$

A method anologous to LAD immediately suggests itself. Setting the time derivative in (9) equal to zero, we define the biharmonic driver method (abbreviated as BID) by

$$\nabla^4\psi^k = H^{k-1} = R_e \nabla\cdot(\vec{V}^{k-1}\nabla^2\psi^{k-1}) \tag{10}$$

All the boundary conditions are incorporated into each BID iteration. For example, the no-slip wall conditions along the bottom wall (j = 1) in Figure 1 are

$$\psi = 0 \quad , \quad \psi_y = 0 \tag{11}$$

This method has not yet been tested, for lack of a direct biharmonic solver, but it would seem to hold great promise.

EXTENSIONS

There appears to be no difficulty in extending these methods to 3D or to the use of primitive (u, v, P) variables, although the latter would seem to offer no advantages either. At present, the extension to compressible flow is prohibited for recirculating flows by the problem of solving the continuity equation, since solutions for mass density do not exist for all velocity fields.

More complete discussions will be found in Roache (1972a).

CONCLUSIONS

Two iterative methods have been presented for solving the steady-state incompressible Navier-Stokes equations. These methods, the numerical Oseen (NOS) and Laplacian driver (LAD) methods are not time-dependent nor even time-like in their iterations. They have been successfully applied to several problems using the stream-function (ψ) and vorticity (ζ) equations, and have performed better than the simplest time-dependent methods. Numerical experiments have shown that the lagging boundary values of ζ slow the convergence more than the lagging nonlinear advection term. Another method, utilizing a direct biharmonic solver, has been suggested to overcome this difficulty. There are fundamental difficulties in applying the methods to compressible flow.

ACKNOWLEDGMENT

The author gratefully acknowledges Professor O. Buneman of Stanford University, who wrote a special version of his Poisson solver used in this work.

REFERENCES

Briley, W. R. (1970), "A Numerical Study of Laminar Separation Bubbles Using the Navier-Stokes Equations," Report J110614-1, United Aircraft Research Laboratories, East Hartford, Connecticut, U.S.A.

Buneman, O. (1969), "A Compact Non-Iterative Poisson Solver," SUIPR Report No. 294, Stanford University, Stanford, California, May 1969.

Davis, R. T. (1972), "Numerical Solution of the Navier-Stokes Equations for Laminar Incompressible Flow Past a Parabola," J. Fluid Mechanics, Vol. 51, Part 3, pp. 417-433.

Dorr, F. W. (1970), "The Direct Solution of the Discrete Poisson Equation on a Rectangle," SIAM Review, Vol. 12, No. 2, pp. 248-263, March 1970.

Roache P. J., (1970), "Sufficiency Conditions for a Commonly Used Downstream Boundary Condition on Stream Function," J. Computational Physics, Vol. 6, No. 2, pp. 317-321.

Roache, P. J. (1971), "A Direct Method for the Discretized Poisson Equation," SC-RR-70-579, Sandia Laboratories, Albuquerque, New Mexico, February 1971. See also Proc. Second Intn'l Conf. on Numerical Methods in Fluid Mechanics, M. Holt, ed., Springer-Verlag, 1971.

Roache, P. J. (1972a), "Finite Difference Methods for the Steady-State Navier-Stokes Equations," SC-RR-72-0419, Sandia Laboratories, Albuquerque, New Mexico, August 1972.

Roache, P. J. (1972b), Computational Fluid Dynamics, to be published.

Roache, P. J. (1972c), "On Artificial Viscosity," to be published in J. Computational Physics.

Schlichting, H., (1968), Boundary Layer Theory, 6th Edition, McGraw-Hill Book Co., Inc., New York.

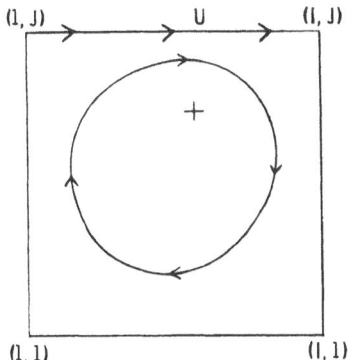

Figure 1. Driven cavity Problem

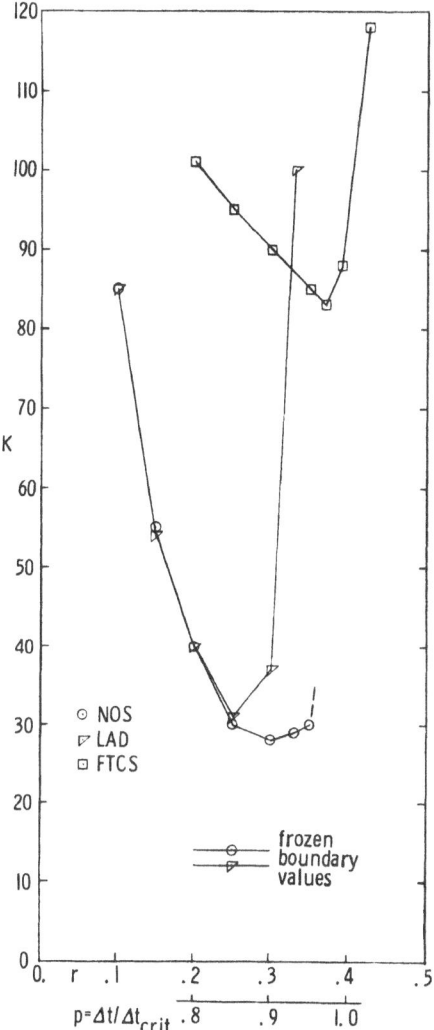

Figure 2. Convergence results for a driven cavity problem, Re = 20, $\Delta x = \Delta y = 1/10$, eqn. (4a) for wall ζ. K is the number of iterations required to satisfy $DQ^K/r < 0.0001$

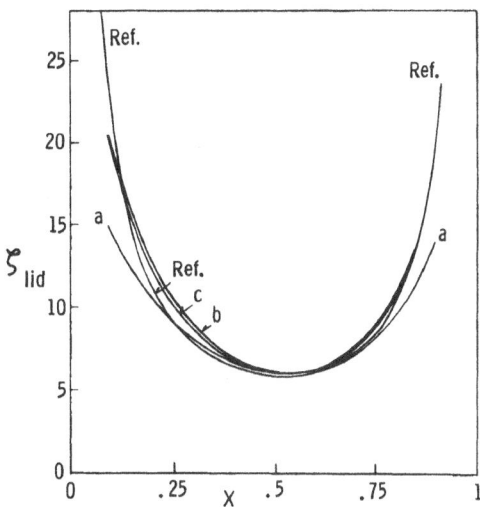

Figure 3. Vorticity distribution along the lid for driven cavity, Re = 20, $\Delta x = \Delta y = 1/10$

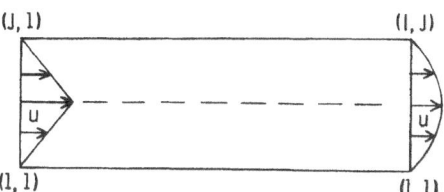

Figure 4. Channel flow problem

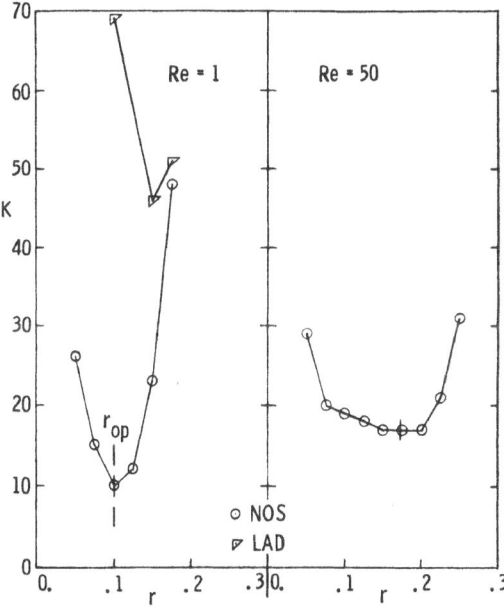

Figure 5. Convergence results for a channel flow problem, Re = 20, $\Delta y = 1/10$, $\Delta x = 1$, eqn. (4b) for wall ζ

A PREDICTOR-CORRECTOR METHOD FOR THREE COORDINATE VISCOUS FLOWS [†]

Stanley G. Rubin

Polytechnic Institute of Brooklyn
Preston R. Bassett Research Laboratory
Farmingdale, New York

I. INTRODUCTION

In a recent paper, Rubin and Lin (1972) have presented a study of a predictor-corrector multi-step finite-difference method. This technique is particularly suited to three-coordinate viscous flows, including transient two-dimensional boundary layer and Navier-Stokes problems, as well as steady three-dimensional flows where diffusion is important in one or two directions. Therefore, boundary region effects and problems with large cross flow diffusion and even cross flow separation are treatable. Rubin and Lin have considered a model Burgers equation in addition to their application to the hypersonic leading-edge region of a cruciform geometry. Imbedded shock formation has been obtained, agreement with experimental data is good and comparisons with previous explicit calculations, requiring in excess of an order-of-magnitude more computer time, are excellent.

The present paper will briefly review the accuracy, stability and consistency of this predictor-corrector (P/C) approach before discussing further applications to a transient model equation and the high-speed flow over a cone at incidence. For the model equation, comparisons with solutions obtained by alternating-direction-implicit (ADI) and Dufort-Frankel (D-F) finite-difference methods are discussed. Specifically, the influence of frequency dependent initial conditions, geometry and non-linearity are evaluated. For the cone geometry, tip or strong interaction solutions are obtained with the single layer analysis of Rudman and Rubin (1968) and Rubin et al. (1969). Shock location, pressure and heat transfer distributions are determined, and "lift-off" of the leeside vortical layer is predicted.

Finally, boundary layer solutions for the weak interaction region downstream are presented. A marching procedure is applicable and therefore unlike three-dimensional boundary layer calculations where similarity is postulated, e.g., Cooke (1966), a complete streamwise plane solution is not required in order to initiate a cross plane calculation. Crossflow separation and vortex formation are obtained, as are surface and boundary layer distributions for all stations from the windward to the leeward plane for angles of incidence up to twice the half-cone angle. Only selected results are depicted herein. Complete details of these three-dimensional analyses and numerical procedures for cone flows will be presented in two papers by Lin and Rubin (1972, 1972a).

[†]This research is sponsored by the Air Force Office of Scientific Research, Office of Aerospace Research, USAF, under Grant No. AFOSR 70-1843 and Modification No. AFOSR 70-1843A, Project No. 9781-01.

II. ACCURACY, STABILITY AND CONSISTENCY

The predictor-corrector finite-difference technique discussed herein has the following distinguishing characteristics: (1) application to three-coordinate viscous flow problems as previously described, where marching procedures are applicable; (2) finite-difference quotients that are always implicit in the surface normal direction but "explicit" in the cross flow direction for steady considerations or along the surface in transient cases. In this way the choice of the normal grid, which in most boundary layer analyses must be small in order to insure acceptable resolution, does not adversely affect the numerical stability. Furthermore, by multiple iteration, crossflow induced instability becomes Reynolds number independent; (3) second-order accuracy and consistency of the difference and differential equations is obtainable. In this regard, multiple iteration and the choice of initial predictor values at each step play important roles; (4) in view of the unidirectional iteration, computer coding is simplified and there is no essential difficulty with non-rectangular geometries and all forms of symmetry boundary conditions; in addition, problems with diffusion in one or two directions are treated in the same manner.

For a model Eq. (1) and with a Crank-Nicolson formulation the following difference Eqs. (2) typify the P/C method.

$$u_t + cu_y + du_z = Re^{-1}(\delta u_{yy} + u_{zz}) \tag{1}$$

$$u_y = (\tfrac{1}{4}\Delta y)\left[u_{I+1,J+1,K} - u_{I+1,J-1,K} + u_{I,J+1,K} - u_{I,J-1,K}\right] \tag{2a}$$

$$u_{yy} = (\tfrac{1}{2}\Delta y^2)\left[u_{I+1,J+1,K} + u_{I+1,J-1,K} - 2u_{I+1,J,K} + u_{I,J+1,K}\right.$$
$$\left. + u_{I,J-1,K} - 2u_{I,J,K}\right] \tag{2b}$$

$$u_z = (\tfrac{1}{4}\Delta z)\left[u^m_{I+1,J,K+1} - u^m_{I+1,J,K-1} + u_{I,J,K+1} - u_{I,J,K-1}\right] \tag{2c}$$

$$u_{zz} = (\tfrac{1}{2}\Delta z^2)\left[u^m_{I+1,J,K+1} + u^m_{I+1,J,K-1} - 2u_{I+1,J,K} + u_{I,J,K+1}\right.$$
$$\left. + u_{I,J,K-1} - 2u_{I,J,K}\right] \tag{2d}$$

$$u_t = (1/\Delta t)\left[u_{I+1,J,K} - u_{I,J,K}\right] \tag{2e}$$

$$u = \tfrac{1}{2}\left[u^m_{I+1,J,K} + u_{I,J,K}\right] \tag{2f}$$

m denotes the iteration number and it is to be understood that the m+1 superscript applies to all non-superscripted I+1 variables. Initial predictor values, i.e., terms with superscript zero, are prescribed by

linear replacement: $u^0_{I+1,J,K} = u_{I,J,K}$ $\qquad$ (3a)

Taylor series to $O(\Delta t^2)$: $u^0_{I+1,J,K} = 2u_{I,J,K} - u_{I-1,J,K}$ $\qquad$ (3b)

Taylor series to $O(\Delta t^3)$: $= 3u_{I,J,K} - 3u_{I-1,J,K} + u_{I-2,J,K}$ $\qquad$ (3c)

Stability

With c,d,Re constant, $\delta=1$, and (2) and (3a) inserted into (1), the following interior point stability conditions have been obtained (see Rubin and Lin (1972)):

$\quad$ m=0,1: If $|d|\Delta z Re \leq 2$, the calculation is unconditionally stable with a viscous dominated cross flow.

m=0: If $|d|\Delta zRe > 2$, the calculation is stable only if
$\Delta t \leq (\Delta z^2 Re)[(|d|\Delta zRe/2)^2-1]^{-1}$

m=1: If $|d|\Delta zRe > 2$, we now require $\Delta t \leq \Delta z/|d|$. There is no longer any Re dependence and a standard CFL condition based solely on crossflow properties results.

m>1: There are only minimal changes from m=1 conditions.

With (3b), Reynolds number independence is achieved without iteration.

Consistency

For fixed $\beta = Re^{-1}\Delta t/\Delta z^2$ and $\Delta t, \Delta y, \Delta z \to 0$, consistency of the difference and differential equations is achieved except when (3a) is applied for the initial predictor value. In this case an error proportional to $\sigma_m = [\beta/(1+\beta)]^{m+1}$ occurs and only with many iterations (m>>1) or small step sizes ($\beta << 1$) is consistency achieved. With (3b) an error of order Δt is incurred (see Section III) and with (3c) true second-order accuracy (Δt^2) results.

Accuracy

The accuracy of the multi-step P/C method is demonstrated with a model non-linear Burgers equation, c=0,d=u,δ=0,Re=1. The solutions are shown on Fig. 1. The present calculations are compared with the exact solution as well as a fully implicit Crank-Nicolson calculation. Stable and highly accurate results are obtained for relatively large β values with only two or three iterations. Iteration is minimized with (3c) as might be expected. Additional non-linear examples are presented by Rubin and Lin (1972) where a more detailed stability and consistency analysis can be found. In the next section the model Eq. (1) and accuracy of solution will be considered further.

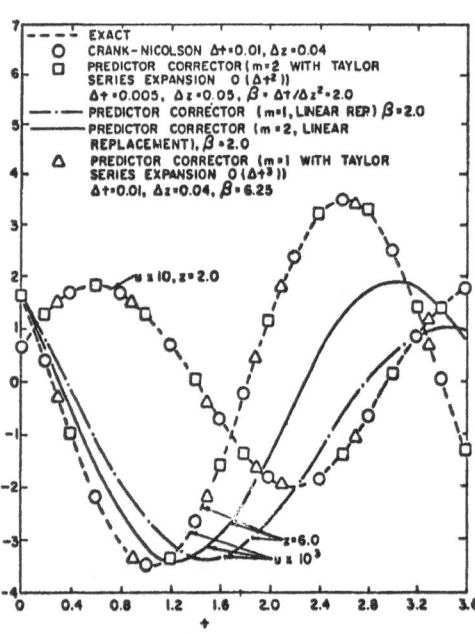

Fig. 1. Solution for Burgers equation

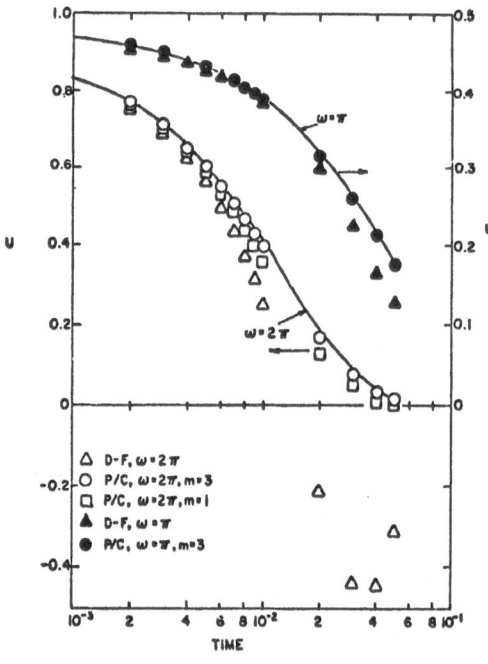

Fig. 2. Frequency dependence β=10,ϵ_o=10^{-2} y=0.2, z=0.3

III.   FREQUENCY DEPENDENT INITIAL CONDITIONS

In certain applications to transient flows where the initial conditions are frequency dependent, the accuracy or "consistency" of a difference scheme, in a practical sense for small but finite $\Delta t, \Delta z$ can depend strongly on the magnitude of the frequency. A specified degree of accuracy may then be achieveable only with a significantly reduced step size. This type of behavior is described in the following example, where it is shown how the multi-step P/C method can ease this difficulty with a minimum number of iterations.

Consider (1) with $c=d=0$; $\delta=Re=1$, so that

$$u_t = u_{yy} + u_{zz} \tag{1a}$$

Boundary and initial conditions are given by

(i) $t=0$; $u=\sin\omega y\sin\omega z$ (therefore, from (1a)

$$u_t = -2\omega^2\sin\omega y\sin\omega z) \tag{4}$$

(ii) $y=z=0$ and $y=z=\pi/\omega$; $u=0$

The exact solution is $u=e^{-2\omega^2 t}\sin\omega y\sin\omega z$ (5)

With the P/C differences (2) and (3b), or a D-F formulation, the governing difference equation describes the modified differential Eq. (6)

$$\hat{u}_t = \hat{u}_{yy} + \hat{u}_{zz} - \epsilon_m\hat{u}_{tt} \tag{6}$$

where m is the iteration number and $\epsilon_0=\beta\Delta t=\Delta t^2/\Delta z^2$. In a D-F approach, with $\Delta y=\Delta z$, $\epsilon_m=2\epsilon_0$ throughout, while with the P/C method $\epsilon_1=(\Delta t/2)/(1+2\beta)$ and $\epsilon_m \ll 1$ with $m \gg 1$.

The solution of (6) with (4) is

$$\tilde{u} = \hat{u}/\sin\omega y\sin\omega z = \exp[-1+(1-8\omega^2\epsilon_m)^{\frac{1}{2}}]t/2\epsilon_m \tag{7}$$

For ϵ_m small but finite, (7) becomes

$$\tilde{u} = e^{-2\omega^2 t} + O(\epsilon_m) \qquad\qquad \text{for } \epsilon_m \ll (8\omega^2)^{-1} \ll 1 \tag{8a}$$

$$\tilde{u} = e^{-4\omega^2 t}(1+2\omega^2 t) \qquad\qquad \text{for } \epsilon_m = (8\omega^2)^{-1} \tag{8b}$$

$$\tilde{u} = e^{-t/2\epsilon_m}\{\cos(8\omega^2\epsilon_m-1)^{\frac{1}{2}}t/2\epsilon_m$$

$$-[(4\omega^2\epsilon_m-1)/(8\omega^2\epsilon_m-1)^{\frac{1}{2}}]\sin(8\omega^2\epsilon_m-1)^{\frac{1}{2}}t/2\epsilon_m\} \quad \text{for } 1 \gg \epsilon_m > (8\omega^2)^{-1} \tag{8c}$$

It is apparent that "consistency" of the difference scheme is not achievable with $\epsilon_m \ll 1$ alone, but in addition we must require $\epsilon_m \ll (8\omega^2)^{-1}$. As the frequency ω increases, this puts a stringent limitation on the size of ϵ_m and therefore, the step size Δt, for a fixed value of Δz; e.g., with $\omega=2\pi$, $\epsilon_m \ll 0.003$.

Solutions for the different frequencies are presented on Fig. 2. Note that the percentage error decreases with decreasing ω and ϵ_m. For the P/C approach, ϵ_m decreases as m increases and for $\epsilon_0=10^{-2}$, $\omega=2\pi$, solutions which differ significantly without iteration, as in the D-F calculation, show a marked improvement with a few iterations. Any specified degree of accuracy can be achieved with additional steps. The D-F calculations with $\omega=2\pi$, $\epsilon_0=0.1$, $\epsilon_m > (8\omega^2)^{-1}$ are in excellent agreement with the oscillatory expression (8c). When (3c) is applied, the difference equations are second order accurate without iteration and frequency effects are less pronounced. The same is true of the ADI technique, which, however, is more sensitive to geometric and non-linear effects to be discussed.

IV. GEOMETRIC AND NON-LINEAR EFFECTS

To further explore the capabilities of the predictor-corrector method and to evaluate the effect of iteration with non-linear terms or symmetry boundary conditions, modified forms of Eq. (1) or boundary conditions (4) are considered. Comparisons are made with DuFort-Frankel and alternating-direction finite-difference calculations.

Equation (1a), having exact solution (5) and boundary conditions (4) ($\omega=\pi$), has been solved for the rectangular geometry inferred in (4) as well as the triangular region bounded by y=0, z=1 and y=z. Across the diagonal the symmetry conditions $u_{I,J,K}=u_{I,K,J}$ are imposed. Secondly, the effect of non-linearity has been considered with the following inhomogeneous model equation for which (5) is once again the exact solution:

$$u_t + u u_y = u_{yy} + u_{zz} + (\omega/2)\sin^2 \omega z \sin 2\omega y$$

Boundary conditions (4) apply to this non-linear case.

The solutions at selected points are tabulated in Table 1. For all of the calculations, the spacial grid is fixed and only the time step changes. It is seen that (1) all of the methods are excellent for the smallest step size considered here, i.e., $\beta=1$, $\Delta t=10^{-4}$, $\Delta z=10^{-2}$; (2) the D-F formulation, as would be anticipated from the consistency analysis, becomes inadequate with the larger values of ε_0; (3) the P/C solutions with two or three iterations are quite acceptable even with the larger β values. Additional iterations are required if further increases in accuracy are desired; (4) the ADI solutions, all obtained without iteration, are excellent for the rectangular and non-linear cases with prescribed boundary values. However, there is a marked drop in accuracy for the triangular geometry and for problems with large timewise and local spacial gradients on the boundaries (not depicted in Table 1). Some form of iteration may improve the situation but previous experience with merged layer calculations (see Rubin and Lin (1972) and Nardo and Cresci (1971)) indicate that similar difficulties occur with more complex non-linear systems and boundary conditions.

TABLE I

z=0.3, y=0.2	$\beta=\Delta t/\Delta z^2$	$\varepsilon_0=\beta\Delta t$	m	t=0.01	t=0.02	t=0.04
Exact				0.3903	0.3204	0.2159
DuFort-Frankel	1	10^{-4}		.3901	.3201	------
	10	10^{-2}		.3837	.2985	.1667
Alternating Direction (ADI)	1	10^{-4}		.3904	.3204	------
	10	10^{-2}		.3903	.3204	.2159
Predictor Corrector (P/C)	1	10^{-4}	2	.3903	.3204	------
	5	2.5×10^{-3}	3	.3901	.3201	.2157
	10	10^{-2}	3	.3892	.3188	.2144
ADI-Nonlinear	1	10^{-4}		.3904	.3204	------
	10	10^{-2}		.3904	.3205	.2160
P/C Nonlinear	1	10^{-4}	2	.3903	.3204	------
	10	10^{-2}	3	.3892	.3188	.2144
ADI-Diagonal	1	10^{-4}		.4723	.4786	------
	10	10^{-2}		.7245	.4448	.3409
P/C-Diagonal	1	10^{-4}	2	.3903	.3204	------
	10	10^{-2}	3	.3892	.3188	.2144

V. CONE FLOWS AND SUMMARY

In this paper we have reviewed and considered in greater detail the merits of a predictor-corrector finite-difference approach to three coordinate viscous flow problems. Analysis of the most rudimentary model equations and comparisons with existing finite-difference methods have demonstrated some of the favorable features of this multi-step technique. However, the most significant test of the P/C formulation is in application to complex problems of current interest. Rubin and Lin (1972) have considered the flow near the leading edge of a cruciform geometry with considerable success. Imbedded shock formation has been predicted, heat transfer comparisons were excellent and the accuracy, stability and consistency of this technique are all compatible with the model equation analysis. In the final portion of this paper we will discuss some recent solutions obtained for cones at incidence in both the tip and downstream regions.

Tip Flow: Merged Layer and Strong Interaction Region

The viscous interaction near the tip of the cone is treated with the single layer analysis of Rubin et al.(1969). To review, the entire disturbed flow including shock wave, boundary layer and inviscid core regions, should they develop, are described by a single uniformly valid set of equations. Marching techniques are applicable and the effects of crosswise diffusion are contained in the governing system so that cross flow separation and imbedded shock formation can be evaluated. Typical solutions, on a cone of half angle $\theta=10°$, for surface pressure and shock location are shown in Fig. 3. Also depicted is the experimental data of Tracy (1963) measured further downstream on the cone. Note that conical flow conditions are achieved surprisingly close to the tip and more rapidly on the windward than on the leeward plane. The heat transfer does not exhibit as rapid an approach to the asymptotic conical values, Fig. 4. Large entropy gradients are predicted for the leeward plane indicative of the diffusion of the vortical singularity into a thin entropy layer. For small angles of incidence α, $\alpha/\theta < 0.8$, the flow in the leeward plane is directed toward the surface while for larger yaw angles, $\alpha/\theta \geq 0.8$, a "lift-off" phenomenon is observed. Crossflow separation is not apparent and therefore does not originate from the tip of the cone. Further details and complete flow profiles for $\alpha/\theta \leq 2$ are presented by Lin and Rubin (1972).

Boundary Layer Calculation

Three-dimensional boundary layer solutions for cones at incidence, $\alpha/\theta \leq 2$, have been obtained with the P/C formulation. With a streamwise marching procedure, and with the inclusion of all pertinent cross diffusion terms, complete boundary layer profiles, surface conditions and streamline behavior from the windward to the leeward plane have been determined. For the first time, to this author's knowledge, cross flow separation, pairwise vortex formation and leeplane solutions have been obtained for $\alpha/\theta \leq 2$.

The heat transfer distributions for $\alpha=4°,16°$, (Fig. 4) are in good agreement with the experimental data of Tracy (1963). On Fig. 5 streamlines projected on the crossflow plane are depicted for $\alpha=8°$ and $20°$. At $20°$ a vortex motion lying completely within the boundary layer and with supersonic cross velocities has been calculated. At $8°$ the results indicate that similarity is not achieved on the leeward side and this signals the failure of any analysis relying on this approximation.

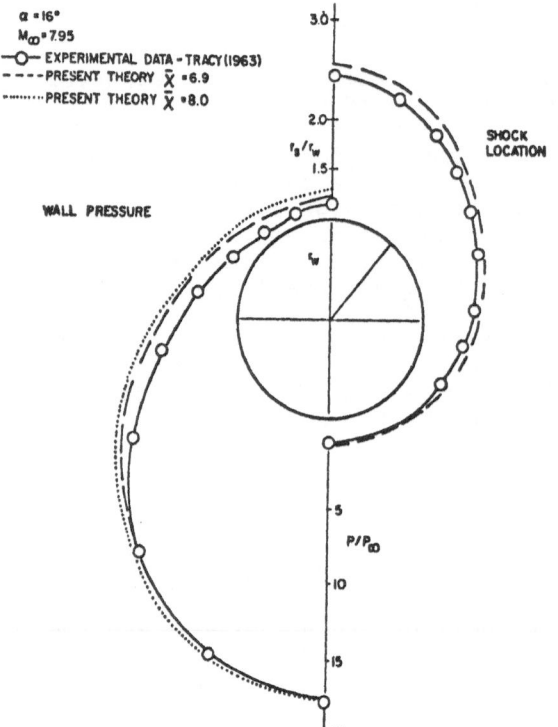

Fig. 3. Tip solution for pressure and shock locations

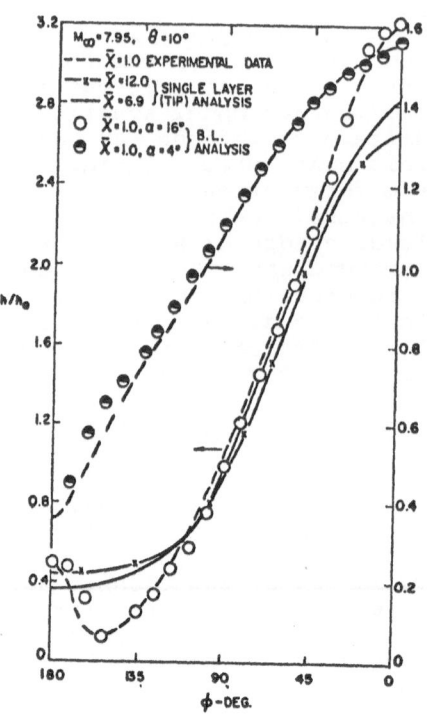

Fig. 4. Heat transfer distribution

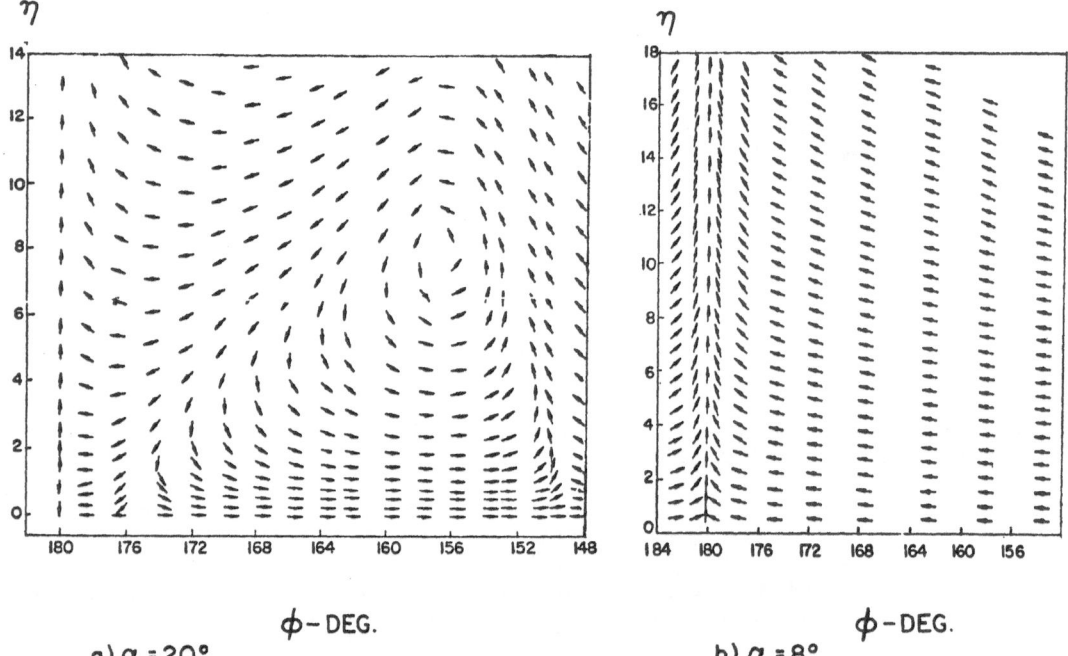

a) $\alpha = 20°$

b) $\alpha = 8°$

Fig. 5. Secondary flow patterns; $M_\infty = 7.95$, $\theta = 10°$

A detailed study of these cone boundary layer calculations are contained in a paper to appear shortly by Lin and Rubin (1972a), where complete solutions for the entire range of incidence angles are presented. The evolution of vortices, crossflow separation, separation line shear layer formation, failure of similarity approximations, symmetry plane solutions, and surface flow patterns are among the topics discussed in this paper.

The author would like to thank Dr. T. Lin and Mr. M. Distefano for their help with the computations and manuscript preparation.

REFERENCES

Cooke, J.C., R.A.E. Tech. Rep. 66347 (1966).
Lin, T.C. and Rubin, S.G., to appear Computers and Fluids (1972).
Lin, T.C. and Rubin, S.G., to appear (1972a).
Nardo, C.T. and Cresci, R.J., J. Comp. Phys., 8, 268-284 (1971).
Rudman, S. and Rubin, S.G., AIAA J., 6, 1883-1890 (1968).
Rubin, S.G. et al., AIAA J., 7, 1744-1751 (1969).
Rubin, S.G. and Lin, T.C., J. Comp. Phys., 339-364 (1972).
Tracy, R.R., GALCIT Memorandum No. 69 (1963).

PROCESSING AND ANALYSIS OF COMPUTATION RESULTS FOR

MULTIDIMENSIONAL PROBLEMS OF AEROHYDRODYNAMICS

V. V. Rusanov

U.S.S.R. Academy of Sciences, Moscow

1. Introduction

This paper is concerned with the questions of processing and representation of numerical information describing functions of several independent variables. In accordance with the accepted tradition we shall call a problem one-, two-, or three-dimensional if its solution depends essentially on one, two, or three spatial variables. If there is time dependence as well, the problem will be called unsteady, otherwise - steady.

One of the specific features of numerical solution of multi-dimensional problems is the large amount of information to be processed in the course of the solution, which exceeds by several orders the amount of output information of real interest to a research worker. This fact was pointed out by John von Neumann as early as 1949 [1] (Published in 1966). For two decades since that time computer development went on rapidly and today much more complicated problems are solved with the help of computers, and as a result the amount of information to be processed has increased tremendously. In this connection a somewhat paradoxical but rather natural situation has arisen: the proportion of effectively useful information has become essentially smaller relative to the total amount of information which is processed in a computer. The question is not about the ratio of the total numbers describing results of a computation to the total numbers with which the calculations are performed; the fact that this ratio is very small ($10^{-6} \div 10^{-7}$ and even smaller) is not a reason for surprise or anxiety. The essence of the matter is that the features of solution of multidimensional problems are not studied in such detail as in one-dimensional problems. The main reason for this is that an effective analysis of functions of several variables cannot be carried out effectively by a human being by means of the same methods which were used for one-variable functions. Firstly, the multivariable functions are more sophisticated and secondly, the representation of multivariable functions becomes more difficult as the number of variables increases.

The present paper contains an attempt to systematize the procedures of processing information in multidimensional problems, and on the basis of available experience discusses its total structure.

2. The General Description of the Numerical Solution of Multidimensional Problems

Let $u(x,t)$ be a solution of the physical problem, which we are interested in. Here $x = \{x_1, x_2,\dots,x_d\}$ are coordinates of a point in d-dimensional space R^d, t is the time, $\bar{u}$ is the vector function with the components $u_k(x,t)$, k = 1,2, 3,\dots,K, representing unknown physical quantities. Let us assume that u is a solution of the following mixed problem in the region G of the space R^d with the boundary Γ and for $t \in [t°,T]$:

$$\frac{\partial u}{\partial t} = L(u,x,t) \tag{1a}$$

$$\Phi[u|_{\Gamma}, x, t] = 0 \tag{1b}$$

$$u(x,t°) = \phi(x) \tag{1c}$$

Here L is some differential operator defined on the functional class $D(R^d)$, Φ
and ϕ are given vector functions.

We shall assume that the stated problem is correct without describing more
exactly the conditions necessary for this.

We shall distinguish between the following two practically important cases.

(I) The original physical process depends essentially on time, and the purpose
of our computation is to determine and analyze the function u(x,t) in the region
G and in some time interval $[t°,T]$. In this case we shall say that the problem I
is unsteady.

(II) The initial physical process does not depend on time and system (I) is
such that its solution u(x,t) has a limit as $t \to \infty$ independent of time, i.e.,

$$\lim_{t \to \infty} u(x,t) = U(x) \tag{2}$$

The purpose of our computation is to determine the function U(x) and to analyze
it. In this case we shall say that the problem of determining the U(x) is steady
and can be solved by a stabilization method.

For the numerical solution of equations (1) it is necessary to approximate
them in a certain way by equations in some discrete space. The basis of the
approximation process consists in assigning to any function v(x,t) some finite
set of parameters p(v) in such a way, that by means of some algorithm $A_\epsilon(p)$
one can reconstruct the value v(x,t) at any point (x,t). In other words, the
set of parameters p(v) together with the decoding algorithm $A_\epsilon(p)$ represent
a table for the function v(x,t) in region G. By using this representation the
system (1) can be replaced by a system of finite equations for a set of parameters.
There exist many methods of tabulation, for example, the assignment of function
values at some fixed points (with instructions for interpolation), the assignment
of coefficients of a Taylor or Fourier series, the assignment of coefficients of a
linear combination of the fixed functional system, etc. The application of each
method depends on the class of the function and additional requirements for its
tabulation.

The representation of a function by its values at points of some fixed mesh is
one of the most universal methods of tabulation. It is the basis of the finite-
difference method for the numerical solution of differential equations. Taking
this as an illustration, let us describe the conversion from the system (1) to
finite difference equations.

Let us introduce in the region G a mesh with the cell dimensions $\Delta t = \tau$,
$\Delta x_\ell = h_\ell$, and write

$$x_\ell^{m_\ell} = x_\ell^0 + m_\ell h_\ell \quad , \quad t^n = t^0 + n\tau \quad .$$

Introducing the generalized index $m = \{m_1, m_2, \ldots, m_d\}$, we write

$$x^m = \{x_1^{m_1}, x_2^{m_2}, \ldots, x_d^{m_d}\}$$

and the coordinates for the mesh points are (x^m, t^n). We suppose that for
$(x^m, t^n) \, \epsilon \Omega = G \times [t°,T]$ the set of indices m is M and for the index n takes
values from 0 to N.

Let u(x,t) be a solution of system (1) belonging to a class D and v(x,t)
be some function from D. Let p(v) be the set of the values of the v(x,t) at
the mesh points, i.e., $p(v) = \{v^{m,n}\}$ where $v^{m,n} = v(x^m, t^n)$ and let the decoding
algorithm $A_\epsilon(p)$ be some interpolation method which leads to the function $\tilde{v}(x,t)$
such that

$$\| v(x,t) - \tilde{v}(x,t) \| < \varepsilon \tag{3}$$

Without specifying the norm in (3), note only that it has to estimate not only the function itself but also its derivatives up to an order equal to the maximum order of the derivatives in (1).

Now substitute the function $\tilde{v}(x,t)$ into (1) and fix the values of the left hand sides of the equations at points $(\hat{x}^s, t^s)$, $s = 1, 2, \ldots, S$ in order to obtain for the parameters $v^{m,n}$ a complete set of equations. In accordance with the specific character of system (1) the approximating system or, in other words, a discrete model of the physical process can be written in the form

$$v^{n+1} = F_\lambda (v^n, v^{n-1}, \ldots, v^{n-n_1}, \hat{x}, \hat{t}) \tag{4}$$

where

$$v^n = \{v^{m,n}\}$$

For an unsteady problem the index n corresponds to a physical time instant t^n, increasing as n increases.

In the case of calculation by the stabilization method the index n does not necessarily correspond to the physical time instant but can be simply the number of the iteration. The result of the calculation, as stated previously, is the limit value of the parameter set

$$V = \{v^m\}_{m \in M} = \lim_{n \to \infty} v^n$$

or, in effect, $V = v^N$.

3. The Processing Algorithms

Using modern computers it is possible to solve very complicated physical problems with high accuracy. However, for full realization of these possibilities we require, besides the calculating algorithm (like that described above or some other one) to organize on the computer a special algorithm for processing of the numerical information obtained. In Fig. 1 the connection between the components of the computing process is represented in diagrammatic form.

The study of the numerical information has two aspects. On the one hand, $\{v^n\}$ is a discrete function, on the other hand, it is the set of parameters defining a continuous function $\tilde{v}(x,t)$. Both types of study are closely connected and are used jointly. One can list the following main steps of the analysis, which are general for both aspects:

1) Determination of extremal values or boundaries of variation of the v^n.

2) Calculation of difference ratios and values of derivatives.

3) Calculation of functions of the quantities v^n and coordinates of the mesh points, in particular, conversion to new independent and dependent variables; after that one can use steps 1) or 2) once more.

Certainly, it is necessary to construct the function $\tilde{v}(x,t)$ to apply steps 1) - 3) to it. This type of processing covers the essentials and we shall consider it in more detail. The conversion from $v^{m,n}$ to $\tilde{v}(x,t)$ is usually carried out on the assumption that the quantities $v^{m,n}$ represent approximations of the values of the function $u(x,t)$ at the mesh points, i.e., for $u^{m,n} = u(x^m, t^n)$. The small difference between $v^{m,n}$ and $u^{m,n}$ is a consequence of the convergence of the process, and using $v^{m,n}$ instead of $u^{m,n}$ is permissible provided that the algorithm $A_\varepsilon(p)$ is valid for the function class to which the $u(x,t)$ belongs.

However, if the definition of convergence does not require close agreement between derivatives, the quantity $\|\tilde{v}(x,t) - \tilde{u}(x,t)\|$ may not be small in the sense of the norm which includes the estimate of derivatives. Moreover, the application of the algorithm $A_\varepsilon(p)$ to the set of the parameters $v^{m,n}$ may yield a function approximating no smooth function at all. In practice this may arise in the case of theoretical convergence of the solution (4) to a smooth function also, at the expense of a round-off disturbance distorting the values $v^{m,n}$.

This situation is closely connected with the incorrectness of the problem of calculating derivatives. Therefore in algorithms for processing a numerical solution it is reasonable to include special means of regularization and smoothing, as for example, the method of least squares.

Suppose that the function $\tilde{v}(x,t)$ has been constructed, i.e., that the algorithm $A_\varepsilon(p)$ has been defined. Then it is possible to analyze the problem as follows, in addition to the procedure outlined above:

4) Determination of extremal points of the components of the functions $\tilde{v}(x,t)$ where first-order derivatives vanish.

5) Calculation of the values of the $\tilde{v}(x,t)$ on the given surface or on another variety in the space (x,t). Change of variables.

6) Calculation of the set of the points (x^*,t^*) at which some scalar function $f(\tilde{v})$ takes the given value f^*.

7) Experience in calculations shows that the method of analysis of functions just outlined, as well as other methods, enable us to study features of the solution and allow us to find rather refined effects [4].

So far while considering the function $\tilde{v}(x,t)$ we assumed it continuous and smooth. However, when problems of gas dynamics with shock waves are solved, the function $\tilde{v}(x,t)$ is only piecewise continuous. If the calculation is carried out by means of a shock-smoothing scheme, then shocks are to be distinguished in the solution. Up to now, this problem remains very difficult as the algorithm of processing is to contain elements of object recognition. One example of such an algorithm is given below.

It should be noted that the above types of analysis of functions represent the lowest levels of processing. The higher ones are to include recognition algorithms on a larger scale, and also automatic classification of new characteristics, i.e., specific capacities of "artificial intelligence". These higher levels have not been realized yet, but evidently their development is necessary for the effective use of computers in calculations of complicated multidimensional problems.

The types of processing mentioned above can be used in greater or lesser sizes, as may be required. In so doing it is reasonable to distinguish the so-called "static" and "dynamic" processing.

The former is used for studying the solution of a steady problem when the limiting function V is at the investigator's disposal for a long time (certainly being stored in some computer medium). In this case a processing time does not play a great role as it is usually much less than the total time of solving the problem on the computer. Besides, it is feasible to process the same material many times for the purpose of extracting more information, or of clearing the questions which arose in the study.

Dynamic processing is used for solving an unsteady problem when it is impossible to store all the information obtained in the course of a calculation. In this case we limit the processing time so that the total time of operation of the calculation process does not exceed admissible bounds.

As has been said, the concept of processing may include the control of a calculating algorithm. A control process includes the following functions:

1) Data gathering and determination of certain criteria of the problem state, data accumulation.

2) Analysis of the data gathered and generation of instructions of the operations required.

3) Control by changing both physical parameters and those of the numerical algorithm.

The last item can be performed either by means of the program automatically, or by means of interactive facilities, if any. In the control process there are also different levels of automation and later on recognition algorithms and "artificial intelligence" may be included. Figure 2 shows the approximate scheme of processing based on the above considerations.

4.   Some Examples of Processing Algorithms

a) Construction of level curves of a two-variable function.

Let $\{f_{m,n}\} = P$ be a set of values of a continuous scalar function, given at the mesh points $x_m = mh_1$, $y_\ell = \ell h_2$. Take for a decoding algorithm $A_\epsilon(P)$ representing linear interpolation inside a primary cell, so that the function $\tilde{f}(x,y)$ can be calculated at any point (x,y).

The problem is the construction of the level curve of the function $\tilde{f}(x,y)$, i.e., the locus of the points satisfying the condition $\tilde{f}(x,y) = f^*$, where f^* is a given value.

The additional requirement is that the level curve can be automatically constructed by means of a plotter. This requirement makes the problem nontrivial.

Indeed, on the basis of $\tilde{f}(x,y)$ it is easy to obtain a set of points of intersection on the level curve with the lines $x = x_m$ and $y = y_\ell$. The main problem is the automatic ordering of these points, which is necessary for using a plotter.

The idea of one of the ordering methods uses the fact that the direction of the tangent to a level curve at any point is defined by the vector $\{\tilde{f}_y, -\tilde{f}_x\}$. Taking this fact into consideration, it proves to be feasible, for each point of intersection of some branch of the level curve with the coordinate line, to determine the point from adjacent points lying exactly on this branch.

If the required point is close to a singularity, where $\tilde{f}_x = \tilde{f}_y = 0$, then the method described is not suitable. In this case a larger number of points is needed to analyze the behavior of the level curve.

The criterion for the "singular" cell may be the condition of alternating signs of the difference $f_{m,\ell} - f^*$ while following the vertices of the mesh cell which contains the point of the level curve. In this case, all the four sides of the cell are intersected by the level curve, and the singular point may lie in the vicinity of the cell.

The singularity is a saddle point or center for the functions usually encountered in gas dynamics. As stated above, to clear up the type exactly it is necessary to use the values of the $f_{m,\ell}$ at the mesh points adjacent to the cell.

To do this, instead of the function $\tilde{f}(x,y)$ consider the function $\tilde{\tilde{f}}(x,y)$ defined in the vicinity of the "singular" cell by quadratic interpolation. When the function $\tilde{\tilde{f}}(x,y)$ has been constructed, the coordinates of the singular point (ξ^*, η^*) can be calculated, and the type of singularity can be determined. If the point $(\xi*,\eta*)$ is outside the cell, then verification is to be repeated for the cell in which the point $(\xi*,\eta*)$ is found. Figure 3 illustrates the scheme of this algorithm and Fig. 4 the use of it for analysis of behavior of the function close to a singularity of the saddle point type.

b) Recognition of shock waves when using shock-smoothing difference schemes for calculating gas flows.

The existence and location of the shocks has to be cleared up by means of special processing algorithms. We shall consider one of these methods for detecting shocks in one dimensional unsteady or two dimensional steady flows.

In [5] it is shown that when calculating a shock wave by the shock-smoothing scheme in the vicinity of the discontinuity, some limiting profile of distribution of gas dynamic parameters is developed, which moves at the wave velocity. Depending on the parameters of the scheme and the shock wave, the profile may or may not be monotone. The profile of characteristic slope λ is formed correspondingly, and it also moves without distortion with the velocities of the discontinuity. It is not difficult to show that the characteristics of the numerical solution will not intersect, but asymptotically approach some line corresponding to the shock trajectory (or shock line in steady flow)(Fig. 5).

Estimates made for the case of a steady supersonic flow by the method of [5] show that the error in the shock positions is of the order of the mesh step and is much less than the width of the zone of wave diffusion.

Figure 6 shows the field of the characteristics coming together in the flow around a blunt body. Many other applications of this method are contained in [4].

REFERENCES

1. John von Neumann. Theory of selfreproducing automates. University of Illinois Press, 1966.
2. A. G. Vitushkin. Bounds of complexity in problems of tabulation. (Russian) Fizmatgiz, Moscow, 1969.
3. A. A. Samarskii. Introduction to the theory of difference schemes. (Russian) NAUKA, Moscow, 1971.
4. A. N. Liubimov, V. V. Rusanov. Flow of gases past blunt bodies. (Russian) NAUKA, Moscow, 1971.
5. V. V. Rusanov. "Non-linear analysis of the shock profile in difference schemes," in Proc. of the Second Int. Conf. on Numerical Methods in Fluid Dynamics, p. 270, Springer-Verlag, 1971.

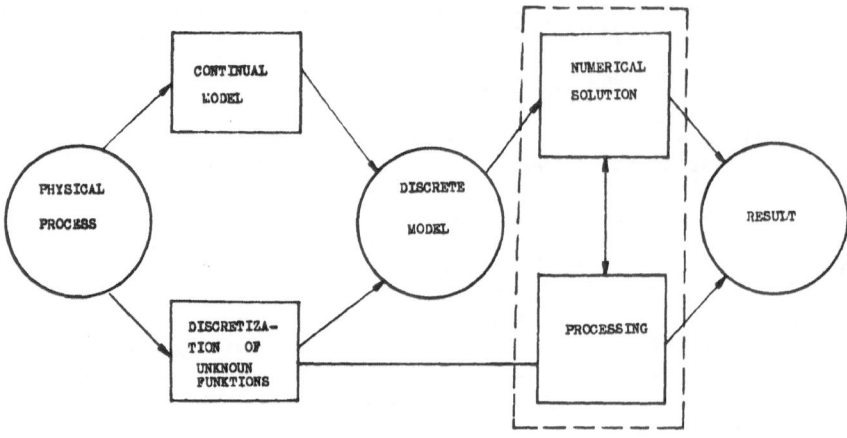

Fig. 1

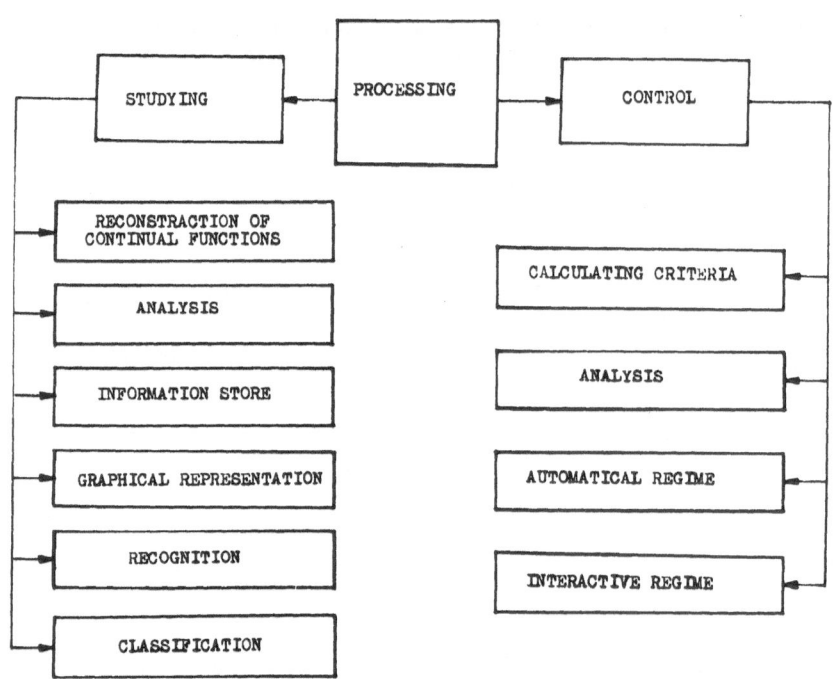

Fig. 2

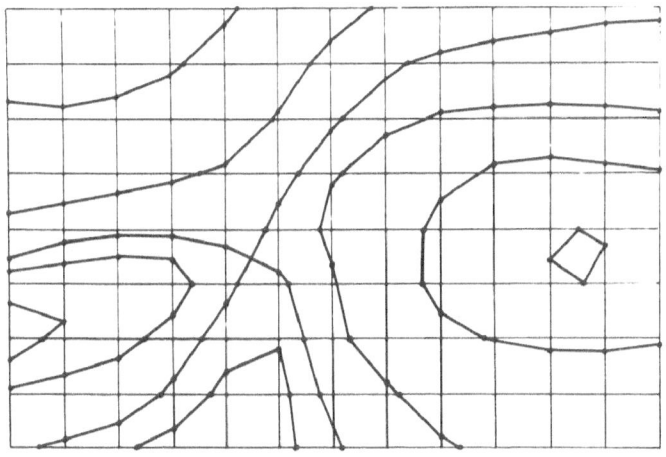

Fig. 3

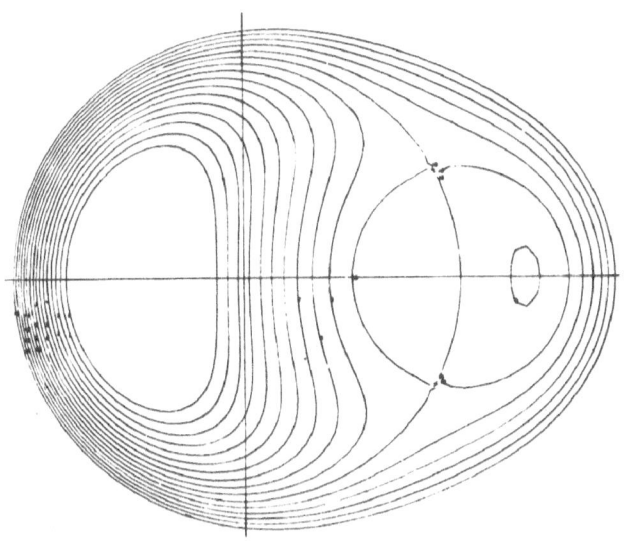

Fig. 4

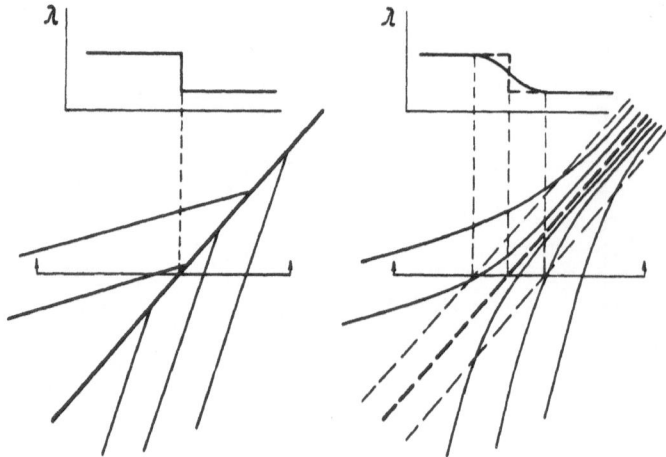

Fig. 5

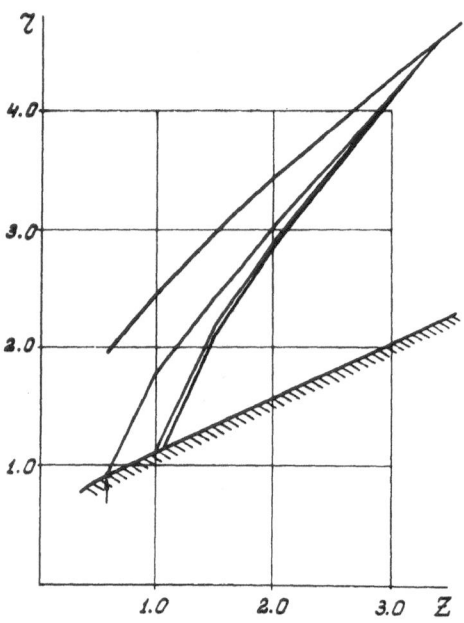

Fig. 6

TOWARDS THE ULTIMATE CONSERVATIVE DIFFERENCE SCHEME
I. THE QUEST OF MONOTONICITY

Bram van Leer

University Observatory, Leiden, Netherlands*

1. THE PROBLEM

The one-dimensional equations of ideal compressible flow are preferably written in conservation form:

$$\frac{\partial w}{\partial t} + \frac{\partial f(w)}{\partial x} = 0. \tag{1}$$

Many interesting flows, notably those containing shocks, can be computed with conservative, dissipative difference schemes based on Eq. (1). In order to analyze or design such schemes it is most practical to start from the single convection equation

$$\frac{\partial w}{\partial t} + a\,\frac{\partial w}{\partial x} = 0. \tag{2}$$

How to make a scheme for Eq. (2) useful to integrate Eq. (1) is explained by Van Leer [4]. The best-known conservative schemes, those of Lax, Godunov and Lax-Wendroff, are based on the cluster of nodal points C1 defined in Fig. 1.

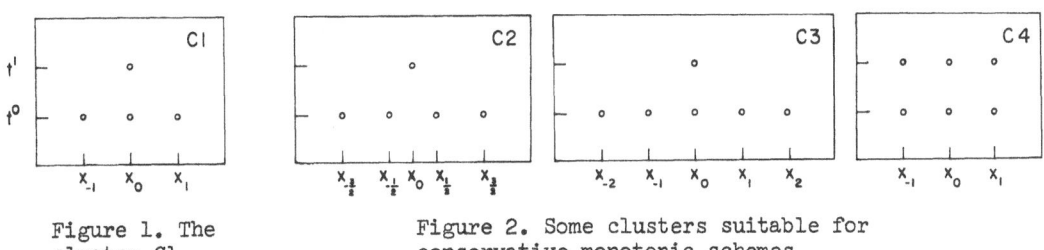

Figure 1. The cluster C1

Figure 2. Some clusters suitable for conservative monotonic schemes

Eq. (2) transforms an initially monotonic distribution such that it remains monotonic at later times; in fact, the original distribution is just shifted over a distance $a \times t$. A difference scheme for Eq. (2) can not deliver this exact solution, except for integer values of $\sigma = a\Delta t/\Delta x$, but it may at least be required to produce <u>monotonic</u> results for all stable values of σ. This requirement is called the monotonicity condition; schemes satisfying it will colloquially be called monotonic schemes.

---

* This work was done while I was on leave of absence at the University of California, Berkeley, as a Miller Fellow.

Godunov [1] tested for monotonicity among the linear schemes for Eq. (2) defined on the cluster Cl. These can be written as

$$w^0 = w_0 - \frac{\sigma}{2}(w_1 - w_{-1}) + \frac{q}{2}(w_1 - 2w_0 + w_{-1}); \qquad (3)$$

the notation is clarified in Table 1.

TABLE 1

Notation used in the grid

symbol	definition
x_0	abcissa of central nodal point in cluster Cl
$x_{\pm 1}$	$x_0 \pm \Delta x$
t^0	initial time level
t^1	advanced time level $= t^0 + \Delta t$
w_0, $w_{\pm 1}$	initial values of w in x_0, $x_{\pm 1}$
w^0	value of w in x_0 at advanced time level
$\Delta_{\frac{1}{2}}$	$w_1 - w_0$
$\Delta_{-\frac{1}{2}}$	$w_0 - w_{-1}$

Any function $q(\sigma)$ defines a scheme; for stability it is required that

$$\sigma^2 \leqq q \leqq 1, \qquad (4)$$

and for monotonicity:

$$|\sigma| \leqq q \leqq 1. \qquad (5)$$

The lower limit in (4) yields the second-order scheme of Lax-Wendroff; any other choice of q results in first-order accuracy. The scheme of Lax, corresponding to the upper limit in (4), is the least accurate of all. Godunov's scheme, corresponding to the lower limit in (5), is the best first-order scheme: it has the smallest q that still guarantees monotonicity. One might also say that this scheme maintains the optimum balance between dissipative and dispersive errors. The dissipation is just strong enough to damp shorter waves before they get to much out of step and show up as oscillations on top of the larger features.

Godunov proved there are no linear second-or-higher-order schemes for Eq. (2) that always preserve monotonicity. Such schemes can only handle very smooth initial values, in which higher derivatives are of minor importance. Whoever wants to pursue unconditional monotonicity must take refuge in nonlinear techniques. In the following, regard a and q as functions of w, their values varying from point to point.

Lax and Wendroff [3] showed how their scheme can be made more dissipative, without affecting its order of accuracy, by increasing $q_{\pm\frac{1}{2}}$ by an amount $\sim \frac{1}{2} |\Delta_{\pm\frac{1}{2}} a| / |a|_{\pm\frac{1}{2}}$. This helps damping numerical oscillations as well as nonlinear instabilities, which often go hand in hand. However, the extra dissipation is not necessarily most effective in places where it is most needed. In consequence, monotonicity is not generally achieved. Harten and Zwas [2] use the expression $|\Delta_{\pm\frac{1}{2}} a| / \max_x |\Delta a|$, which gives stronger but still rather arbitrary damping.

In the present paper I shall show that unconditional monotonicity involves the use of the ratio $\Delta_{+\frac{1}{2}} a / \Delta_{-\frac{1}{2}} a$ in the dissipation coefficient. Unfortunately, a scheme based on C1 that includes this expression can never be conservative; see Van Leer [4, Sec. 2.2]. To me, this represents the first valid reason to step up to a larger number of points in the basic cluster. Suitable five- and six-point clusters are shown in Fig. 2. Having no results ready for these more elaborate clusters, I shall occupy myself with the Lax-Wendroff scheme in the further sections of this paper.

2. A NONLINEAR DEVICE

Regard (2) again as a linear equation. To fix our thoughts, let a be positive. I shall start from the Lax-Wendroff scheme, raising q above the value σ^2 as the initial values of w lack smoothness. In this way a nonlinear but monotonic scheme is obtained for a linear equation. For the sake of clarity, the derivation of this scheme given below follows a geometrical rather than an algebraic line of reasoning.

Fig. 3 shows the full range of circumstances under which the ordinary Lax-Wendroff scheme may produce non-monotonic results; the limiting cases of just monotonic behaviour are included. The smoothness of initial-value triplets (w_{-1}, w_0, w_1) is most conveniently characterized by the ratio

$$\zeta = \frac{2 \Delta_{-\frac{1}{2}} w}{\Delta_{\frac{1}{2}} w - \Delta_{-\frac{1}{2}} w}. \tag{6}$$

In the sequence of Fig. 3 this expression runs from -1 to +1.

In each graph a parabola is drawn (thin curve marked LW) through the points $(x = -\Delta x, w = w_{-1})$, $(x = 0, w = w_0)$ and $(x = \Delta x, w = w_1)$. The new value w^0, as computed with the Lax-Wendroff scheme, can be found on this parabola at $x = -\sigma \Delta x$, somewhere in the left mesh. An explanation of this quadratic interpolation routine can be found in Van Leer [4, Sec. 3.5]. For $-1 < \zeta < 1$ the parabola LW undershoots the dotted line LL which indicates the Lower of the Levels $w = w_{-1}$ and $w = w_0$. For

$\zeta = \pm 1$, LW does not undershoot: it is just tangent to LL. For $|\zeta| > 1$, not illustrated here, there is no more danger of undershooting.

In most circumstances where the Lax-Wendroff scheme predicts a new value lower than any of the initial values, there is no reason at all to believe that such lower values really occur in the exact solution. Thus the undershooting value does not represent a gain in accuracy but merely shows the low compliance of the interpolation curve LW. If monotonicity is to be achieved, LW must be replaced in the danger zone $|\zeta| < 1$ by a different curve.

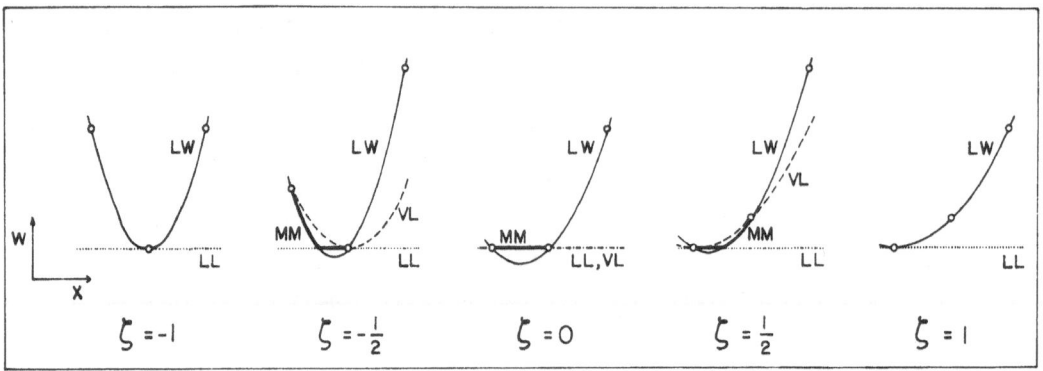

Figure 3. The danger zone of ζ. The choice of concave initial values $(\Delta_{\frac{1}{2}}w - \Delta_{-\frac{1}{2}}w > 0)$ is arbitrary; a similar sequence exists for convex data

The curve closest to LW that still preserves monotonicity is obtained by cutting LW short with LL, whenever LW dips under LL. The result is the heavy curve MM (for Marginal Monotonicity) with the kink; it is indicated only in the left mesh. Using MM in a scheme, however, is halting between two opinions: if the curve LW is not trusted, why retain even a part of it? With the application to nonlinear equations in mind, it is better anyway to choose a curve lying safely above MM, preferably one without a kink. For $\zeta = \pm 1$, the curve must coincide with LW in order to achieve a smooth transition between the "emergency" scheme and the ordinary Lax-Wendroff scheme. The simplest possible replacement curve is a parabola that runs through w_{-1} and w_0 but is not forced through w_1; instead, it is chosen tangent to the line LL. This parabola is the broken curve VL in Fig. 3; for $\zeta = 0$, VL degenerates into LL itself.

The emergency scheme corresponding to the interpolation curve VL can be written as

$$
\begin{aligned}
w^0 - w_0 &= -\sigma^2 \Delta_{-\frac{1}{2}}w, & -1 < \zeta < 0; \\
w^0 - w_0 &= -\sigma(2 - \sigma)\Delta_{-\frac{1}{2}}w, & 0 \le \zeta < 1.
\end{aligned}
\tag{7}
$$

It is easily verified that this scheme can be obtained from the Lax-Wendroff scheme by adding to q the term

$$\sigma(1 - \sigma)(1 - |\zeta|), \qquad |\zeta| < 1. \tag{8}$$

Note that this expression is positive definite, hence damping is indeed increased.

3. A NUMERICAL EXAMPLE

Now assume that a depends linearly on w, such as in the quadratic conservation law

$$\frac{\partial w}{\partial t} + \frac{\partial \left(\frac{1}{2}w^2\right)}{\partial x} = 0, \tag{9}$$

where $a = w$. I redefine ζ as

$$\zeta = \frac{2\Delta_{-\frac{1}{2}}a}{\Delta_{\frac{1}{2}}a - \Delta_{-\frac{1}{2}}a}, \tag{10}$$

which for Eq. (9) gives the same result as definition (6). For the numerical experiment reported below I added the non-conservative ζ-term (8) to the conservative Lax-Wendroff scheme for Eq. (9). In (8) I substituted for σ the local value σ_0. With the resulting scheme a rarefaction wave[1] was followed which at $t = 0$ was defined by the values

$$
\begin{aligned}
w_m &= \frac{1}{3}, & m &\leqq 25; \\
w_m &= \frac{2}{3}, & m &= 26; \\
w_m &= 1, & m &\geqq 27.
\end{aligned} \tag{11}
$$

The results after 24 time-steps, with $\sigma = \frac{1}{2}$, are shown in Fig. 4 together with the results for the unmodified Lax-Wendroff scheme and with the exact solution. Both schemes show a starting error in the position of the wave of about $-0.3\Delta x$. At the head of the wave, the schemes give almost identical results; at the foot, the monotonic scheme is clearly superior. The representation of the foot can be made even more acute by basing the scheme on a curve that fits MM more tightly than VL does. Furthermore, it is worth considering a reduction of the value of q for $-3 < \zeta < -1$, when the parabola LW undershoots LL in the right mesh and should not yet be taken to seriously. With the scheme thus modified, the head of the rarefaction wave will turn out sharper too.

---

[1] I did not make the obvious shock wave computation, as the non-conservative scheme would have produced meaningless results.

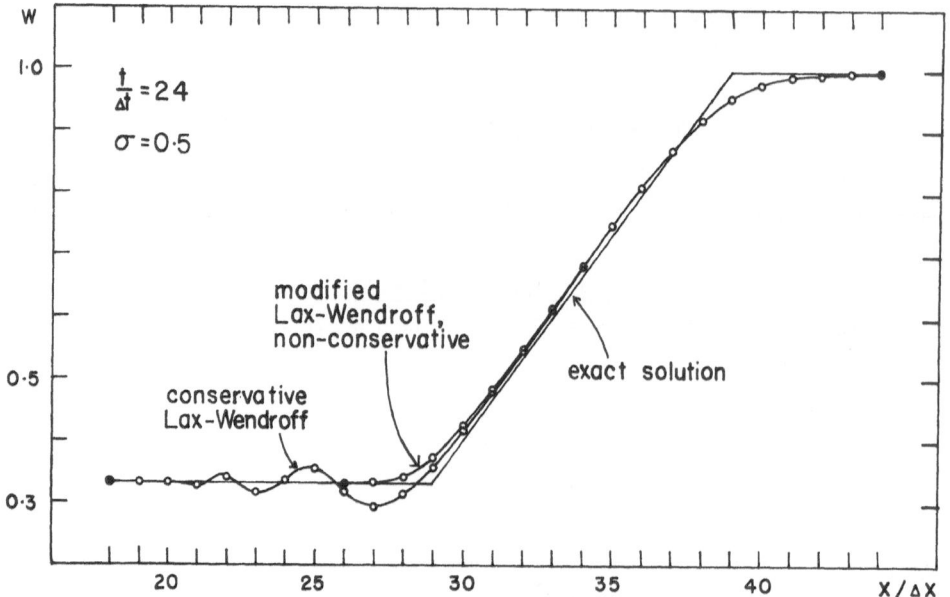

Figure 4. Numerical representation of a rarefaction wave. Beyond the black dots, the numerical results differ less than 0.0001 from the exact solution

The above results clearly demonstrate the usefulness of the smoothness monitor (10) in constructing a monotonic scheme for a single nonlinear conservation law. The main point is now to design a monotonic scheme that is also conservative. The final step is applying it to a nonlinear system of conservation laws, in particular, the equations of ideal compressible flow.

REFERENCES

1. Godunov, S. K., <u>Mat. Sb.</u> 47, 271 (1959); also Cornell Aeronautical Lab. Transl.

2. Harten, A., and Zwas, G., to appear in <u>J. Computational Phys.</u>

3. Lax, P. D., and Wendroff, B., <u>Comm. Pure Appl. Math.</u> 13, 217 (1960)

4. Van Leer, B., "A Coice of Difference Schemes for Ideal Compressible Flow", thesis, University Observatory, Leiden, Netherlands (1970)

POSITIVE CONSERVATIVE SECOND AND HIGHER ORDER DIFFERENCE SCHEMES FOR THE EQUATIONS OF FLUID DYNAMICS

Joseph P. Wright, Associate
Paul Weidlinger, Consulting Engineer
New York, New York

I. INTRODUCTION

In many continuum mechanics problems, there are quantities, such as the mass and energy density functions, which are inherently positive. In addition, the equations of continuum mechanics are based on certain conservation principles. There have been numerous papers which have presented difference approximations for maintaining these conservation principles exactly, and numerous (albeit fewer) papers on difference schemes for preserving positivity everywhere at all times. There is, of course, no strict necessity for using a difference scheme which insures positivity or conserves exactly. Theoretically, it is sufficient if the scheme is stable and consistent with the basic equations so that these properties are attained in the limit of sufficiently small space and time intervals. However, it is intuitively appealing to include as many features of a problem as possible.

The convection (or transport) terms in the Eulerian form of the basic equations often cause difficulty during computations. The main point of this paper is to show how a particular difference operator, which approximates the convection terms, arises naturally from transformations which explicitly account for certain positive functions. This operator can be used to construct positive, conservative, second and higher order difference schemes for problems involving convection terms. Because positivity is taken into account, a natural norm is available and stability of the schemes can be proved readily.

II. MASS CONSERVATION

The main idea of the difference scheme can be seen by considering the Eulerian form of the conservation of mass equation in one dimension

$$\rho_t + (u\rho)_x = 0, \tag{1}$$

where ρ is the mass density and u is the particle velocity. By letting $\rho = f^2$, Eq. (1) can be transformed into the equation

$$f_t + uf_x + \frac{1}{2} fu_x = 0. \tag{2}$$

A consistent, second order accurate, finite difference approximation to Eq. (2), which is obtained by using central differences in time and space, is given by

$$f_i^{n+1} - f_i^n + \frac{\Delta t}{2\Delta x} Q_i (u^{n+\frac{1}{2}}, f^{n+\frac{1}{2}}) = 0, \tag{3}$$

where

$$Q_i (u,f) = u_{i+\frac{1}{2}} f_{i+1} - u_{i-\frac{1}{2}} f_{i-1} , \tag{4}$$

and

$$f_i^{n+\frac{1}{2}} = \frac{1}{2} (f_i^{n+1} + f_i^n). \tag{5}$$

Unconditional stability of this approximation for any size time step Δt can be proved by the energy method by multiplying Eq. (3) by the quantity $(f_i^{n+1} + f_i^n)$ and summing over all values of i. We observe that, except for boundary terms, the Q-operator has the property that

$$\sum_i f_i Q_i (u,f) = 0, \tag{6}$$

and therefore, it is a simple matter to show that

$$\sum_i (f_i^{n+1})^2 = \sum_i (f_i^n)^2. \tag{7}$$

Thus, Eq. (3) has the property that mass is conserved in the same sense as in Eq. (1).

III. SOLUTION OF THE TRIDIAGONAL SYSTEM

Equation (3) represents a tridiagonal system of linear equations which can be written in the form

$$a_i f_{i-1}^{n+1} + b_i f_i^{n+1} + c_i f_{i+1}^{n+1} = d_i^n \tag{8}$$

where
$$a_i = - \frac{\Delta t}{4\Delta x} u_{i-\frac{1}{2}}^{n+\frac{1}{2}} , \quad b_i = 1, \quad c_i = -a_{i+1} ,$$

and
$$d_i^n = f_i^n - \frac{\Delta t}{4\Delta x} Q_i (u^{n+\frac{1}{2}}, f^n).$$

Such a system can be solved efficiently, without loss of accuracy, by means of Gaussian elimination without pivoting if the system is diagonally dominant (Forsythe and Moler, p. 117). This reduces to the condition

$$\max \left| u_{i-\frac{1}{2}} \right| \Delta t < 2\Delta x. \tag{9}$$

Thus, for reasons of numerical accuracy, it is advisable to restrict the time step Δt by a condition which is similar to the Courant-Friedrichs-Lewy stability condition for hyperbolic equations. This problem of numerical accuracy has been observed in computation.

If time steps which violate Eq. (9) are used, then pivoting can be employed in order to avoid loss of accuracy in solving the system of equations. However, this will also increase the solution time. No condition similar to Eq. (9) is known to the author in this case. However, it should be pointed out that, in the limit $\Delta t \to \infty$, Eq. (8) approaches a system which is badly conditioned; in particular, if the system is of odd order it becomes singular and if the system is of even order the set of space points with even subscripts becomes independent of the set with odd subscripts. This difficulty is a reflection of the fact that, in the limit $\Delta t \to \infty$, Eq. (3) represents a difference approximation to the ill-posed problem of solving a first order differential equation with boundary conditions prescribed at both ends of an interval. This seems to indicate that there is a practical limit on the time step whether or not pivoting is used.

IV. ENERGY AND MOMENTUM CONSERVATION

Consider the conservation of energy equation

$$\varepsilon_t + (u\varepsilon)_x + pu_x = 0, \tag{10}$$

where p is the pressure and ε is the internal energy per unit volume. Physically meaningful internal energy densities can always be assumed to be positive so that a transformation similar to that for the mass density can be used. By letting $\varepsilon = g^2$, Eq. (10) can be transformed into the equation

$$g_t + ug_x + \frac{1}{2} gu_x + \frac{p}{2g} u_x = 0. \tag{11}$$

By using $\rho = f^2$, and the conservation of mass equation, the conservation of momentum equation,

$$(\rho u)_t + (\rho u^2)_x + p_x = 0, \tag{12}$$

can be transformed into the equation

$$h_t + uh_x + \frac{1}{2} hu_x + \frac{1}{f} p_x = 0, \tag{13}$$

where h = fu.

V. MULTIDIMENSIONAL PROBLEMS

The multidimensional generalization of Eqs. (3), (11) and (13) can be derived by using the same transformations as above and these equations can be written in the form

$$F_t + (u \cdot \nabla) F + \frac{1}{2} F(\nabla \cdot u) + R(F) = 0, \tag{14}$$

where $F = (\sqrt{\rho}, \sqrt{\varepsilon}, \sqrt{\rho}\, u)$ and $R(F) = (0, p(\nabla \cdot u), \nabla p/\sqrt{\rho})$. If Eq. (14) is written in finite difference form, using central differences in space and time, with

$$F_i^{n+\frac{1}{2}} \equiv \frac{1}{2} (F_i^{n+1} + F_i^n),$$

it can be shown that the Q-operator results, that the difference equations satisfy all of the conservation laws upon which the continuum equations are based and that the appropriate functions are positive. The difficulty with this is that the resulting difference scheme is nonlinear and highly implicit.

An efficient method for integrating the multidimensional equations can be developed by using operator splitting techniques, based on alternating direction and fractional step methods, (Yanenko). As an example, consider a problem involving two space dimensions. A scheme which retains all of the desired conservation and positive properties is to split the operator in alternating directions so that two one dimensional problems are obtained. A further splitting of each one dimensional problem is then used whereby the convection terms are integrated during the first fractional step and the remaining terms are integrated in the second step. The major problem at this point is to develop an efficient method for solving the equations in this second step since quite complicated equations of state may be involved. The velocities which are used in the Q-operator during the first step need not be obtained by iteration since the accuracy of the scheme can be maintained by predicting them by means of an explicit scheme.

Significant improvement over second order schemes was reported by Roberts and Weiss (1966), who investigated the use of fourth order schemes for convection problems. A fourth order operator, $\bar{Q}$, can be constructed by combining the Q-operator, Eq. (4), with the Q-operator of space increment $2\Delta x$,

$$Q_i^{'} (u,f) = u_{i+1} f_{i+2} - u_{i-1} f_{i-2}$$

so that the second order error terms cancel. Thus,

$$\bar{Q}_i (u,f) = \frac{1}{3} [\frac{2}{\Delta x} Q_i (u,f) - \frac{1}{4\Delta x} Q_i^{'} (u,f)].$$

The use of $\bar{Q}$ in place of Q in Eq. (3) results in a scheme which is fourth order in space but still second order in time. Although the system of linear equations analogous to Eq. (8) is quintdiagonal, the remarks made concerning numerical accuracy with and without pivoting still apply.

Although the use of transformations of dependent variables can lead to incorrect solutions when discontinuities, such as shocks, are part of the solution (Kasahara and Houghton), the transformations used above should cause no such difficulties since the proper physical quantities are conserved (Courant, p. 150).

One method for obtaining a positive type scheme is to use a difference approximation in which values at the new time are expressed in terms of values at the old time using a linear combination with positive coefficients. Lax (1961) has shown that, for hyperbolic systems, this method is limited to first order accuracy (except for special cases). Transformations such as those suggested in this paper provide a method for circumventing this result.

There are many fields of study in which conservative, positive schemes are potentially useful. Arakawa (1968) has discussed the need for developing schemes with properties such as this for use in weather prediction and Lathrop (1969) has considered the problem of radiation transport. It is also possible that some of the ideas presented here could be used for solving the Boltzmann equation.

The Q-operator has been successfully used in calculating two dimensional transient and steady-state thermal convection in the presence of a magnetic field (in process of publication). It was in this application that the problem of numerical accuracy for large Δt, mentioned in Section III was first observed.

It should be remarked that this method has not been tried on a problem in which strong shocks form. Whether or not the numerical difficulties which usually accompany a nondissipative scheme for this kind of problem are controlled, so that acceptable solutions are obtained, is not known at this time.

REFERENCES

1. Arakawa, A., "Numerical Simulation of Large-Scale Atmospheric Motions", published in SIAM-AMS Proceedings, Vol. II, pp. 24-40, A.M.S. Publishers, 1970.

2. Courant, R., "Methods of Mathematical Physics", Vol. II, Interscience Publishers, 1962.

3. Forsythe, G.E. and Moler, C.B., "Computer Solution of Linear Algebraic Systems", Prentice-Hall, Inc., 1967.

4. Kasahara, A. and Houghton, D.D., "An Example of Nonunique, Discontinuous Solutions in Fluid Dynamics", J. of Computational Physics, Vol. 4, pp. 377-388, 1969.

5. Lax, P., "On the Stability of Difference Approximations to
    Solutions of Hyperbolic Equations with Variable Coefficients",
    Comm. of Pure and Appl. Math., Vol. 14, pp. 497-520, 1961.

6. Lathrop, K.D., "Spatial Differencing of the Transport Equation:
    Positivity versus Accuracy", J. of Computational Physics,
    Vol. 4, pp. 475-498, 1969.

7. Roberts, K.V. and Weiss, N.O., "Convective Difference Schemes",
    Math. of Computation, Vol. 20, No. 94, pp. 272-299, 1966.

8. Yanenko, N.N., "The Method of Fractional Steps for the Numerical
    Solution of the Problems of Mechanics of Continuous Media",
    published in Fluid Dynamics Transactions, Vol. 4, pp. 135-147,
    Inst. of Fundamental Technical Research, Polish Acad. of Sci.,
    Warsaw, 1969.

SCHEMAS NUMERIQUES INVARIANTS DE GROUPE POUR
LES EQUATIONS DE LA DYNAMIQUE DE GAZ

N.N. Yanenko et Yu.I. Chokin

(Centre de Calcul, Novosibirsk, U.R.S.S.)

1. Dans les travaux $[1]$, $[2]$, les auteurs ont formulé les conditions qui garantissent l'invariance de groupe des schémas numériques précis au 1e ordre. Dans ce cas la forme parabolique de la première approximation différentielle (p.a.d.) admet le même groupe de transformations que le système d'équations original de la dynamique de gaz.

Dans le présent travail on étudie les propriétés dissipatives des schémas numériques invariants.

2. Pour le système d'équations de la dynamique de gaz en coordonnées lagrangiennes

$$W_t = f_x \qquad (1)$$

on considère le schéma numérique

$$\frac{\Delta_o W^n(x)}{\tau} = \frac{\Delta_1 + \Delta_{-1}}{2h} f^n(x) + \frac{1}{h} \frac{1}{2} \left[\Omega(x + \frac{h}{2})\Delta_1 - \Omega(x - \frac{h}{2}) \Delta_{-1} \right] W^n(x) \quad , \qquad (2)$$

où

$$W = \begin{pmatrix} u \\ v \\ E \end{pmatrix} \qquad\qquad f = \begin{pmatrix} -p \\ u \\ -up \end{pmatrix}$$

u - vitesse, v - volume spécifique, p - pression, $E = \varepsilon + \frac{1}{2} u^2$, ε - énergie spécifique interne, $t = n\tau$, τ , h - pas du réseau dans la direction des axes t, x, $\kappa = \tau/h = \text{const}$,

T_o - opérateur de translation sur t, T_1 - opérateur de translation sur x,

$T_{-1} = T_1^{-1}$, E - opérateur d'identité, $\Omega = \|\omega_{ij}\|_1^3$, $|\omega_{ij}| = O(\tau)$,

$\Delta_o = T_o - E$, $\Delta_1 = T_1 - E$, $\Delta_{-1} = E - T_{-1}$.

D'après [1] le schéma numérique (2) admet le même groupe de transformations que le système d'équations (1), si

$$\frac{\partial \omega_{ij}}{\partial t} = \frac{\partial \omega_{ij}}{\partial x} = \frac{\partial \omega_{ij}}{\partial u} = 0 \,, \quad \frac{\partial}{\partial u}\, N_{1x} = 0 \,, \quad \frac{\partial}{\partial u}\, N_{2x} = 0 \,, \quad \frac{\partial}{\partial u}\, N_{3x} = N_{1x} \quad ,$$

$$(3)$$

$$\left\{ u_x \frac{\partial}{\partial u_x} + v_x \frac{\partial}{\partial v_x} + p_x \frac{\partial}{\partial p_x} + 2u_{xx} \frac{\partial}{\partial u_{xx}} + 2v_{xx} \frac{\partial}{\partial v_{xx}} + 2p_{xx} \frac{\partial}{\partial p_{xx}} \right\} N_{\alpha x} = N_{\alpha x}$$

et si de plus, les $N_{\alpha x}$ sont indépendantes des dérivées par rapport à t des fonctions u, v, p.

Ici on a

$$N = \begin{pmatrix} N_1 \\ N_2 \\ N_3 \end{pmatrix} = C\, W_x \,, \qquad C = \Omega - \frac{\tau}{2}\, A^2 \quad , \qquad A = \frac{df}{dw} \quad ,$$

C — matrice des termes visqueux de la forme parabolique de p.a.d. :

$$W_t = f_x + (C\, W_x)_x \,.$$

L'analyse de la stabilité du schéma numérique (2) par la méthode de p.a.d. conduit aux inégalités suivantes :

$$\frac{\tau}{2}\, a^2 \leqslant \beta_i \leqslant \frac{h^2}{2\tau} \qquad (i = 1, 2, 3) \qquad ,$$

où $a^2 = pp_\varepsilon - p_v$, β_i — valeurs propres de la matrice Ω .

3. Parmi les schémas numériques invariants on peut distinguer deux classes. La première classe possède la propriété K, la deuxième est caractérisée par l'absence de la "diffusion de masse", c.-à-d. la condition $N_{2x} = 0$ est vérifiée (cf. [1]).

Si le schéma numérique (2) possède la propriété K et $N_{2x} = 0$ alors la viscosité d'approximation intervient dans la p.a.d. de manière additive, comme c'est le cas pour la viscosité physique, et la matrice C s'écrit dans ce cas sous la forme

$$C = \begin{pmatrix} \nu_{11} - up_\varepsilon \nu_{13} & \nu_{12} + p_v \nu_{13} & p_\varepsilon \nu_{13} \\ 0 & 0 & 0 \\ u\nu_{11} - u^2 p_\varepsilon \nu_{13} & u\nu_{12} + up_v \nu_{13} & up_\varepsilon \nu_{13} \end{pmatrix}$$

où

$$\frac{\partial \nu_{1j}}{\partial t} = \frac{\partial \nu_{1j}}{\partial x} = \frac{\partial \nu_{1j}}{\partial u} = 0 \quad (j = 1, 2, 3)$$

et, de plus, les ν_{1j} ne dépendent pas des dérivées par rapport à t des

fonctions u, v, p .

4. Pour simplifier l'exposé considérons le mouvement d'une onde de choc se

propageant avec la vitesse constante D dans un gaz polytrope. Alors toutes les

caractéristiques de l'écoulement dépendent de la variable $S = x - Dt$. Soit

$$p(+\infty) = p_0 \quad , \quad p(-\infty) = p_1 \quad , \quad v(+\infty) = v_0 \quad , \quad v(-\infty) = v_1 \quad ,$$

$$u(+\infty) = u_\phi = 0, \quad u(-\infty) = u_1 \quad , \quad q(+\infty) = q(-\infty) = 0 \quad ,$$

où

$$q = -\mu_0 h u_x - \xi_0 h v_x - \eta_0 h p_x \quad ,$$

μ_0 , ξ_0 , η_0 - fonctions qui ne dépendent pas de t, x, u ni de

u_t , v_t , p_t .

Si

$$\mu_0 = \bar{\mu}_0 / v \quad , \quad \bar{\mu}_0 = \text{const} \quad , \quad \xi_0 = \eta_0 = 0 \quad ,$$

alors la largeur effective de la zone de transition correspondant à l'onde de choc

est définie par la formule

$$\Delta S = \frac{2 \bar{\mu}_0 h}{\Delta v (\gamma + 1) D} \quad \ln \frac{v - v_1}{v_0 - v} \Bigg|_{\tilde{v}}^{\overset{\approx}{v}}$$

où $\tilde{v}$, $\overset{\approx}{v}$ sont certaines valeurs de v telles que

$$\mid v_1 - \tilde{v} \mid < \delta \qquad , \qquad \mid v_0 - \overset{\approx}{v} \mid < \delta \qquad ,$$

avec δ suffisamment petit.

Si l'on fait diminuer $\overline{\mu}_0$, alors la zone de transition devient plus étroite.

Les fig. 1 - 4 montrent l'effet de la non-invariance du schéma numérique sur les résultats du calcul. Pour le schéma (2) on a choisi la matrice

$$\overline{C} = \begin{pmatrix} \mu_0 h & \xi_0 h & \eta_0 h \\ 0 & 0 & 0 \\ u\mu_0 h & u\xi_0 h & u\eta_0 h \end{pmatrix} \quad , \overline{C}\,\overline{W}_x = C\,W_x , \overline{W} = \begin{pmatrix} u \\ v \\ p \end{pmatrix} \qquad (4)$$

La solution numérique correspond au cas où $D = 1$ et les valeurs initiales :

$$u = 1 \quad , v = 1 \quad , p = 1,5714 \quad , x \leqslant 0 \quad ,$$
$$u = o \quad , v = 2 \quad , p = o,5714 \quad , x > 0 \quad .$$

Les fig. 1 - 2 représentent les profils de volume spécifique v et de vitesse u obtenus d'après le schéma numérique (2) avec la matrice C de la forme (4), où $\mu_0 = \mu_0^1 = \dfrac{h^2 |u_x p_x|}{\delta + h^2 |u_x p_x|}$, $\xi_0 = \eta_0 = 0$ dans les deux espaces : 1) t, x, u, v, p, 2) $t' = t$, $x' = x + Ut$, $u' = u + U$, $v' = v$, $p' = p$, $U = $ const.

Dans ce cas le schéma numérique est invariant, et les profils v, v' et u, u' coïncident compte tenu de la relation $u' = u + U$, ce qu'on voit sur les fig. 1 - 2 . Les fig. 3 - 4 représentent les profils de volume spécifique et de vitesse obtenus d'après le schéma numérique (2) non-invariant ($\mu_0 = u\,\mu_0^1$, $\xi_0 = \eta_0 = 0$) dans les deux mêmes espaces. Les fig. 3 - 4 montrent que la non-invariance du schéma entraîne un écart pour les profils correspondants. Dans ce cas les résultats de calcul dépendent essentiellement de U.

5. Pour étudier plus en détail les propriétés dissipatives du schéma numérique (2) on s'est servi de la deuxième approximation différentielle. Dans ce cas on ajoute à la matrice des termes visqueux une matrice C' telle que la matrice

C + C' soit positive.

L'analyse a montré (et les calculs l'ont confirmé) que dans le cas $\xi_0 > 0$
les profils obtenus numériquement deviennent oscillatoires. La viscosité associée
à la valeur $\eta_0 > 0$ fait augmenter ou diminuer la stabilité selon le signe de
l'expression

$$2\, a^2(a_x)^2 \; + \; \frac{3(\gamma - 1)}{\nu}\; (a^2 \; - \frac{h^2}{\tau^2})p_x.$$

Dans le cas $\xi_0 = \eta_0 = 0$ la condition $\mu_0\, h + \frac{\tau^2}{6}\; \frac{2(\gamma - 1)}{\nu}\; a^2 u_x < 0$
est nécessaire pour la stabilité du schéma.

6. Les fig. 5 - 9 représentent les solutions numériques qui correspondent
aux schémas invariants.

Les fig. 5 - 6 montrent les profils de volume spécifique et de vitesse
pour l'onde de choc stationnaire obtenus d'après le schéma (2) avec $\mu_0 = \overline{\mu}_0 / \nu$,
$\overline{\mu}_0$ = const   , $\xi_0 = \eta_0 = 0$ et différentes valeurs de κ et $\overline{\mu}_0$: 1) $\kappa = 0,1$;
$\overline{\mu}_0 = 1$; 2) $\kappa = 0,4$; $\overline{\mu}_0 = 0,5$; 3) $\kappa = 0,4$; $\overline{\mu}_0 = 1$.

Sur la fig. 7 on voit le profil de pression pour le problème de Cauchy à
valeurs initiales discontinues avec

$$\mu_0 = 0,5\; \frac{h^2|\,u_x p_x\,|}{0,001 + h^2|u_x p_x|} \qquad\qquad (5)$$

Les fig. 8 - 9 représentent les profils de vitesse et de volume spécifique
pour l'interaction de l'onde de choc avec la surface de contact obtenus à partir du
schéma (2) avec le coefficient de viscosité (5).

Les résultats de calcul ont montré :

1) si κ est suffisamment petit le schéma devient instable,

2) on peut diminuer la largeur effective de la zone de transition jusqu'à
deux-trois intervalles,

3) les solutions numériques correspondant aux schémas invariants n'oscillent
que derrière le front de l'onde de choc, elles sont monotones devant le front de
l'onde.

179

La classification de groupe peut être faite d'une manière analogue pour les schémas de haute précision.

Nous allons donner des exemples des schémas numériques de haute précision (de l'ordre 1e , 2e , 3e) pour une équation modèle (cf. aussi [3]) :

$$u_t = \alpha\, y\, u_x - \alpha\, x\, u_y \qquad (6)$$

avec les conditions initiales

$$u(o, x, y) = \begin{cases} 1 - \dfrac{1}{u_o} \sqrt{(x-a)^2 + (y-b)^2} & , \quad (x-a)^2 + (y-b)^2 \leqslant u_o^2 \quad , \\[4mm] 0 & , \quad (x-a)^2 + (y-b)^2 > u_o \end{cases}$$

Le problème (6)-(7) décrit la rotation du cône axisymmétrique de hauteur 1 avec le cercle de rayon u_o comme base (le centre de la base est situé, au moment t = o, au point (a,b)), autour de l'origine des coordonnées avec la période $2\pi / \alpha$. On a posé dans les calculs $u_o = 5$, a = b=2, $\alpha = \dfrac{\pi}{2}$.

L'équation (6) admet la transformation de rotation.

On a construit des schémas numériques approchant l'équation (6), invariants ou non par rapport à la transformation de rotation.

Les fig. 10-15 représentent les courbes de niveau u(t, x, y) = const. (const = 0,2 ; 0,4 ; 0,6 ; 0,8 ; 1) : 1) de la solution exacte (les cercles) et de la solution numérique (la courbe 1 correspond à la solution du schéma numérique non-invariant, la courbe 2 correspond à la solution du schéma invariant). Sur les fig. 10-11 on voit les résultats numériques pour les schémas de 1e et 3e ordre de précision. Sur les fig. 12-15 sont données les solutions des schémas de l'ordre 2 pour les moments 2, 4, 6, 8.

Les fig. 16-17 représentent une section de la solution numérique pour les moments t = 2 et t = 6 par le plan x = -2 .

Bibliographie

[1] N.N. Yanenko, Yu.I.Chokin.   Sur la classification de groupe des schémas
    numériques pour le système d'équations planes de la dynamique de gaz.
    Sb. "Nekotorye problemy matematiki i mekhaniki". L., I970.

[2] N.N.Yanenko, Yu.I.Shokin.   On the group classifications of difference schemes
    of equations in gas dynamics. Lect. Notes in Phys., 8 (I97I), 3 - 17.

[3] S.Z. Burstein, A.A.Mirin.   Third order difference methods for hyperbolic
    equations. J. Comput. Phys., 5(I970), 547-571.

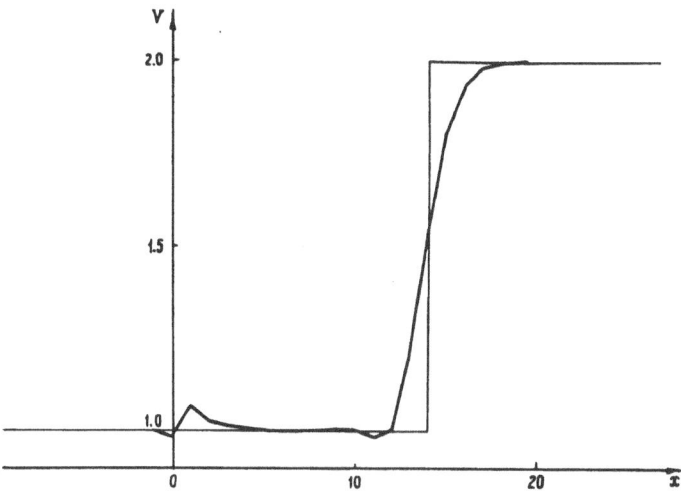

Puc. 1

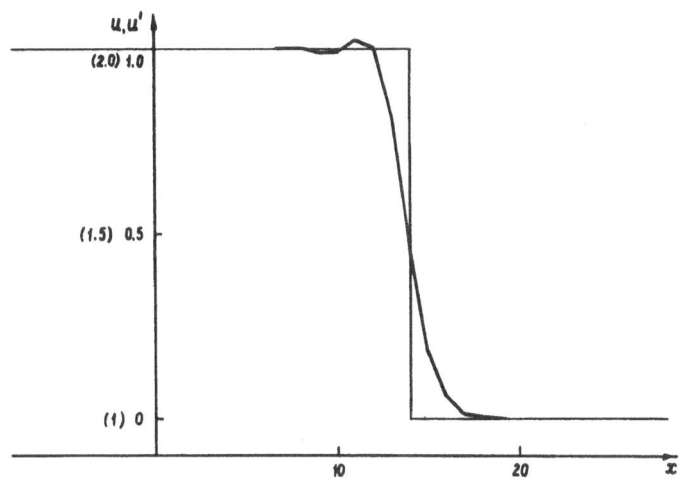

Puc. 2

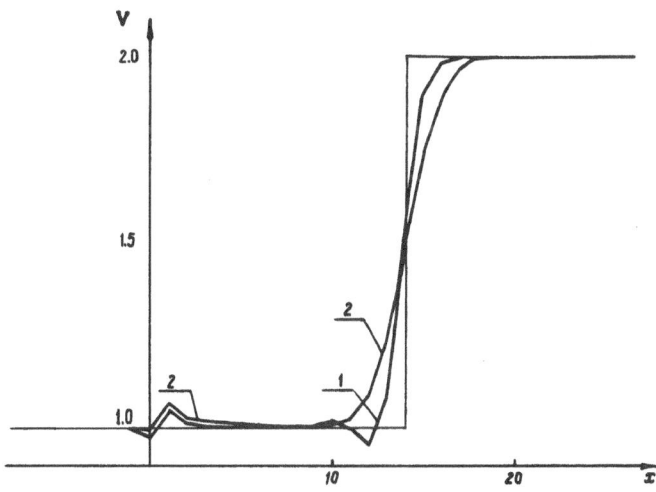

Puc. 3

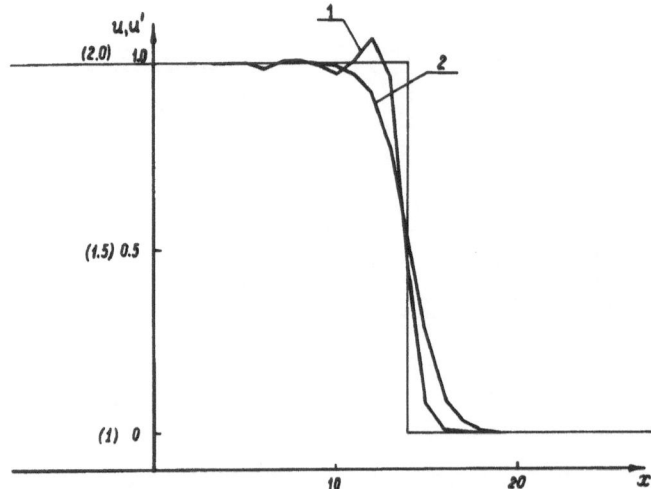

Рис. 4

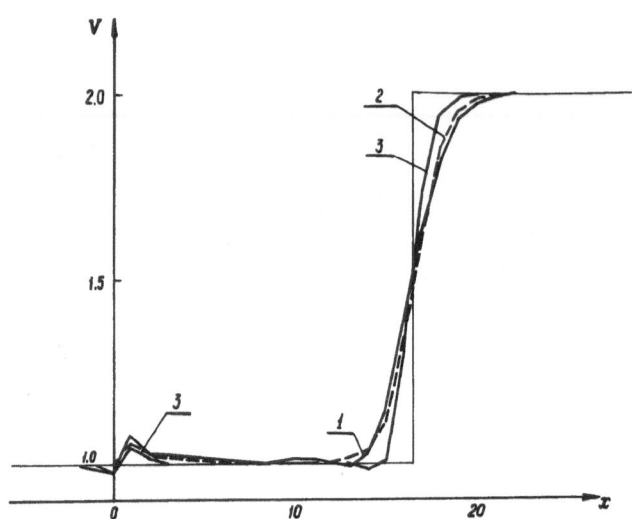

Рис. 5

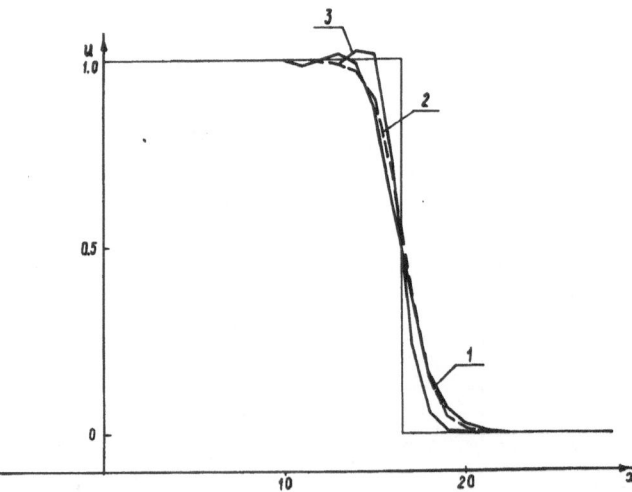

Рис. 6

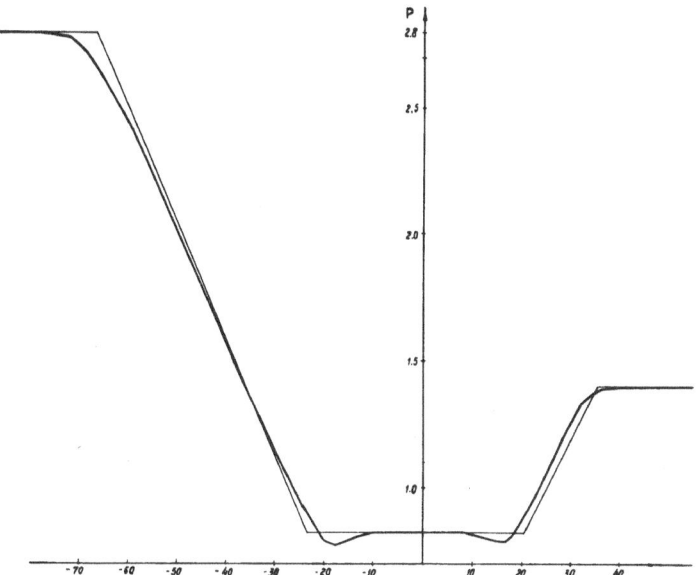

Рис. 7

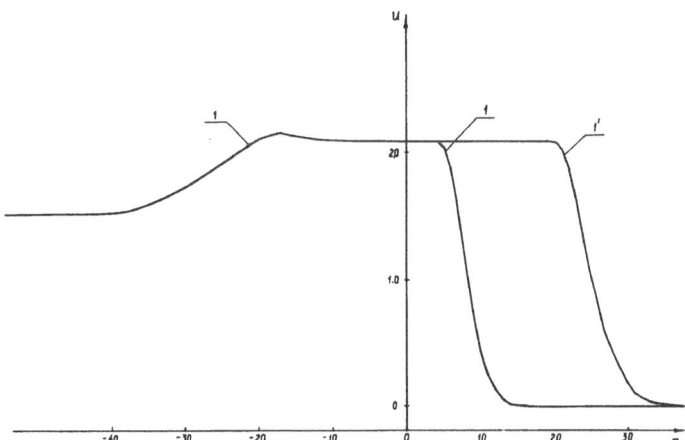

Рис. 8

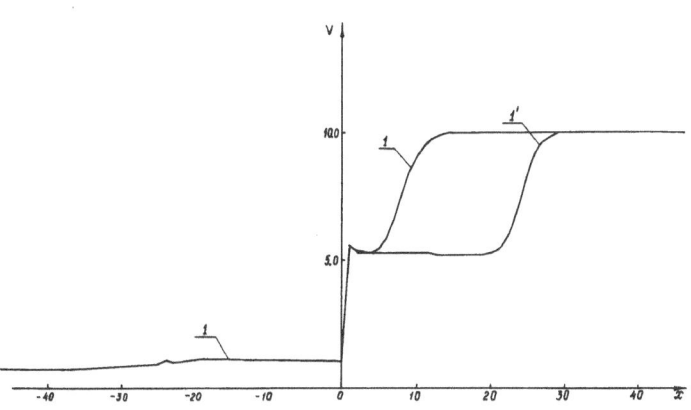

Рис. 9

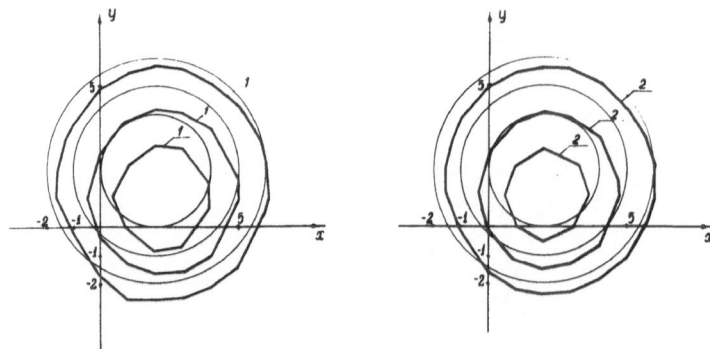

Рис. 10

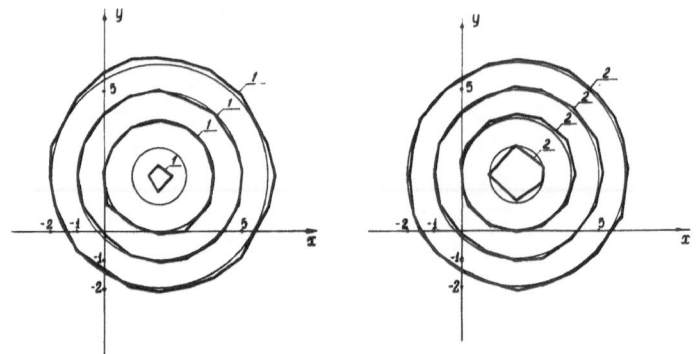

Рис. 11

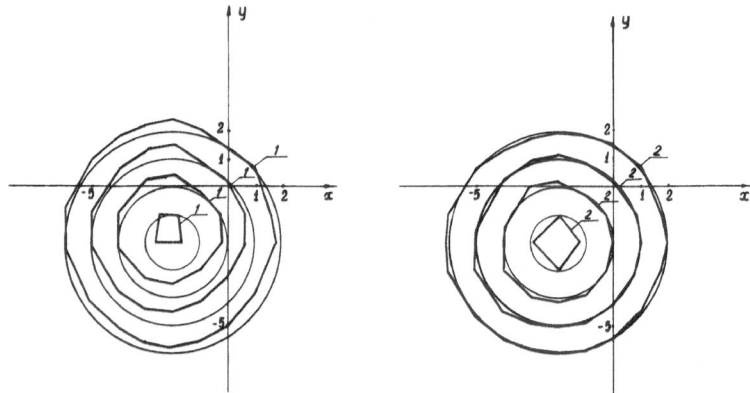

Рис. 12

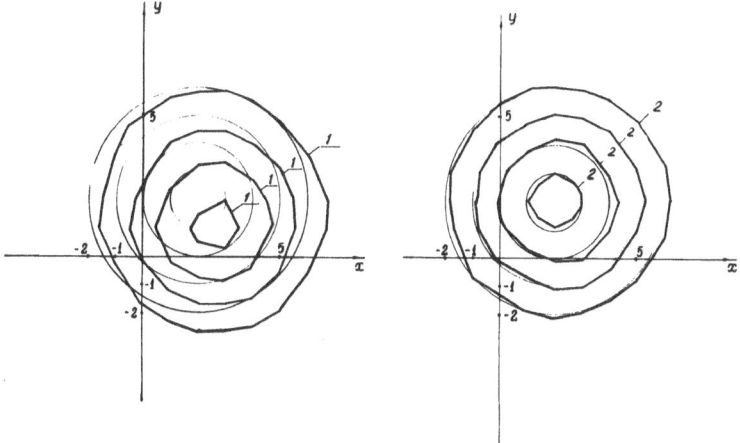

Рис. 13

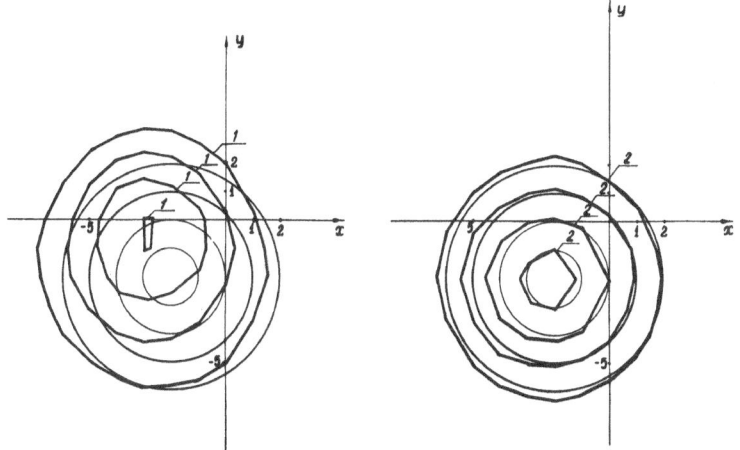

Рис. 14

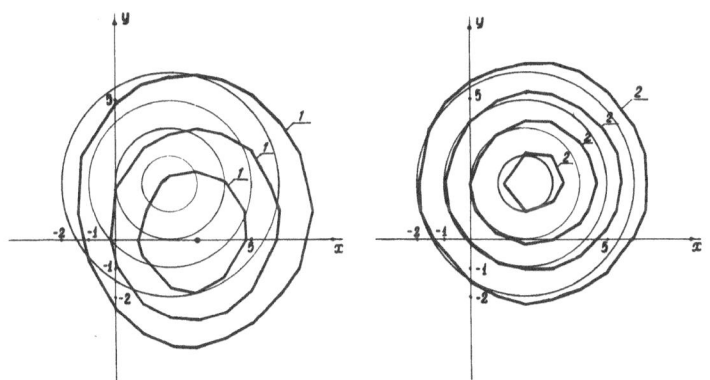

Рис. 15

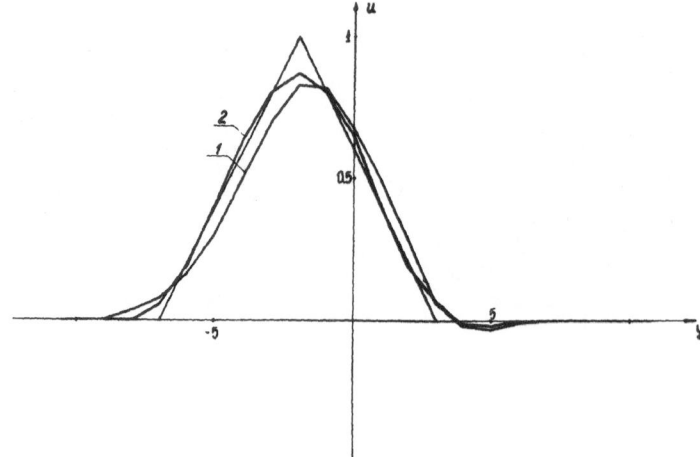

Puc. 16

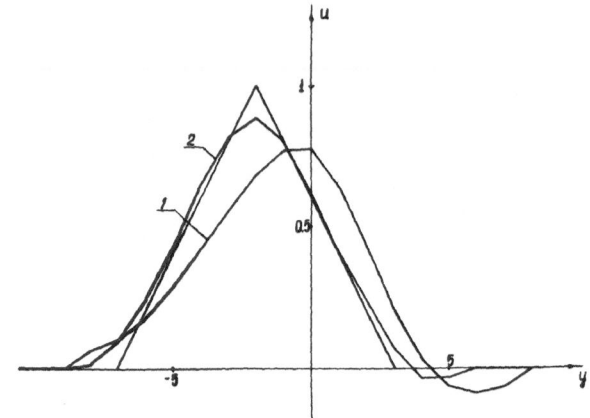

Puc. 17

Lecture Notes in Physics

Bisher erschienen / Already published

Vol. 1: J. C. Erdmann, Wärmeleitung in Kristallen, theoretische Grundlagen und fortgeschrittene experimentelle Methoden. 1969. DM 20,–

Vol. 2: K. Hepp, Théorie de la renormalisation. 1969. DM 18,–

Vol. 3: A. Martin, Scattering Theory: Unitarity, Analyticity and Crossing. 1969. DM 16,–

Vol. 4: G. Ludwig, Deutung des Begriffs physikalische Theorie und axiomatische Grundlegung der Hilbertraumstruktur der Quantenmechanik durch Hauptsätze des Messens. 1970. DM 28,–

Vol. 5: M. Schaaf, The Reduction of the Product of Two Irreducible Unitary Representations of the Proper Orthochronous Quantummechanical Poincaré Group. 1970. DM 16,–

Vol. 6: Group Representations in Mathematics and Physics. Edited by V. Bargmann. 1970. DM 24,–

Vol. 7: R. Balescu, J. L. Lebowitz, I. Prigogine, P. Résibois, Z. W. Salsburg, Lectures in Statistical Physics. 1971. DM 18,–

Vol. 8: Proceedings of the Second International Conference on Numerical Methods in Fluid Dynamics. Edited by M. Holt. 1971. DM 28,–

Vol. 9: D. W. Robinson, The Thermodynamic Pressure in Quantum Statistical Mechanics. 1971. DM 16,–

Vol. 10: J. M. Stewart, Non-Equilibrium Relativistic Kinetic Theory. 1971. DM 16,–

Vol. 11: O. Steinmann, Perturbation Expansions in Axiomatic Field Theory. 1971. DM 16,–

Vol. 12: Statistical Models and Turbulence. Edited by M. Rosenblatt and C. Van Atta. 1972. DM 28,–

Vol. 13: M. Ryan, Hamiltonian Cosmology. 1972. DM 18,–

Vol. 14: Methods of Local and Global Differential Geometry in General Relativity. Edited by D. Farnsworth, J. Fink, J. Porter and A. Thompson. 1972. DM 18,–

Vol. 15: M. Fierz, Vorlesungen zur Entwicklungsgeschichte der Mechanik. 1972. DM 16,–

Vol. 16: H.-O. Georgii, Phasenübergang 1. Art bei Gittergasmodellen. 1972. DM 18,–

Vol. 17: Strong Interaction Physics. Edited by W. Rühl and A. Vancura. 1973. DM 28,–

Vol. 18: Proceedings of the Third International Conference on Numerical Methods in Fluid Mechanics, Vol. I. Edited by H. Cabannes and R. Temam. 1973. DM 18,–

Vol. 19: Proceedings of the Third International Conference on Numerical Methods in Fluid Mechanics, Vol. II. Edited by H. Cabannes and R. Temam. 1973. DM 26,–

Beschaffenheit der Manuskripte
Die Manuskripte werden photomechanisch vervielfältigt; sie müssen daher in sauberer Schreibmaschinenschrift mit ausreichend großer Type geschrieben sein. Handschriftliche Formeln bitte nur mit schwarzer Tusche eintragen. Notwendige Korrekturen sind bei dem bereits geschriebenen Text entweder durch Überkleben des alten Textes vorzunehmen oder aber müssen die zu korrigierenden Stellen mit weißem Korrekturlack abgedeckt werden. Die reproduktionsfähigen Abbildungen (in Originalgröße) sollen in den Text eingeklebt werden. Falls das Manuskript oder Teile desselben neu geschrieben werden müssen, ist der Verlag bereit, dem Autor bei Erscheinen seines Bandes einen angemessenen Betrag zu zahlen. Die Autoren erhalten 50 Freiexemplare.

Zur Erreichung eines möglichst optimalen Reproduktionsergebnisses ist es erwünscht, daß bei der vorgesehenen Verkleinerung der Manuskripte der Text auf einer Seite in der Breite möglichst 18 cm und in der Höhe 26,5 cm nicht überschreitet. Entsprechende Satzspiegelvordrucke werden vom Verlag gern auf Anforderung zur Verfügung gestellt.

Manuskripte, in englischer, deutscher oder französischer Sprache abgefaßt, sind einzureichen bei: Springer-Verlag, 6900 Heidelberg, Postfach 1780.

Cette série a pour but de donner des informations rapides, de niveau élevé, sur des développements récents en physique, aussi bien dans la recherche que dans l'enseignement supérieur. On prévoit de publier.

1. des versions préliminaires de travaux originaux et de monographies

2. des cours spéciaux portant sur un domaine nouveau ou sur des aspects nouveaux de domaines classiques

3. des rapports de séminaires

4. des conférences faites lors de congrès ou de colloques

En outre il est prévu de publier dans cette série, si la demande le justifie, des rapports de séminaires et des cours multicopiés ailleurs mais déjà épuisés.

Dans l'intérêt d'une diffusion rapide, les contributions auront souvent un caractère provisoire; le cas échéant, les démonstrations ne seront données que dans les grandes lignes. Les travaux présentés pourront également paraître ailleurs. Une réserve suffisante d'exemplaires sera toujours disponible. En permettant aux personnes intéressées d'être informées plus rapidement, les éditeurs Springer espèrent, par cette série de «prépublications», rendre d'appréciables services aux instituts de physique. Les annonces dans les revues spécialisées, les inscriptions aux catalogues et les copyrights rendront plus facile aux bibliothèques la tâche de réunir une documentation complète.

Présentation des manuscrits
Les manuscrits, étant reproduits par procédé photomécanique, doivent être soigneusement dactylographiés type assez grand. Il est recommandé d'écrire à l'encre de Chine noire les formules non dactylographiées. Les corrections nécessaires doivent être effectuées soit par collage du nouveau texte sur l'ancien soit en recouvrant les endroits à corriger par du vernis correcteur blanc. Les illustrations; en dimension originale, préparées pour reproduction sont à insérer dans le texte. S'il s'avère nécessaire d'écrire de nouveau le manuscrit, soit complètement, soit en partie, la maison d'édition se déclare prête à verser à l'auteur, lors de la parution du volume, le montant des frais correspondants. Les auteurs recoivent 50 exemplaires gratuits.

Pour obtenir une reproduction optimale il est désirable que le texte dactylographié sur une page ne dépasse pas 26,5 cm en hauteur et 18 cm en largeur. Sur demande la maison d'edition met à la disposition des auteurs du papier spécialement préparé.

Les manuscrits en anglais, allemand ou français peuvent être adressés à Springer-Verlag, 6900 Heidelberg, Postfach 1780.

Springer-Verlag, D-1000 Berlin 33, Heidelberger Platz 3
Springer-Verlag, D-6900 Heidelberg 1, Neuenheimer Landstraße 28-30
Springer-Verlag, 175 Fifth Avenue, New York, NY 10010/USA